RESEARCH FOR THE GLOBAL GOOD

SUPPORTING A BETTER WORLD FOR ALL

RESEARCH FOR THE GLOBAL GOOD

SUPPORTING A BETTER WORLD FOR ALL

Daniel D. Watch
PERKINS
+WILL

images
Publishing

Published in Australia in 2010 by
The Images Publishing Group Pty Ltd
ABN 89 059 734 431
6 Bastow Place, Mulgrave, Victoria 3170, Australia
Tel: +61 3 9561 5544 Fax: +61 3 9561 4860
books@imagespublishing.com
www.imagespublishing.com

The Images Publishing Group Reference Number: 905

A CiP entry for this title is available from the National Library of Australia.

Editor: Chris Wyness

Designed by Mimi Day, Perkins+Will

Production by The Graphic Image Studio Pty Ltd, Mulgrave, Australia
www.tgis.com.au

Pre-publishing services by United Graphic Pte Ltd, Singapore

Printed on 157 gsm NeoMatte FSC paper by Everbest Printing Co. Ltd., in Hong Kong/China

CONTENTS

ACKNOWLEDGEMENTS

There are many people who helped with developing this book but I would like to start by thanking two very important people in my life, my two oldest daughters, Megan, and Kalie, both of whom worked very hard doing research and developing graphics for the book. I also thank my wife, Terrie, and two youngest daughters, Quinn and Lucie, who were very supportive and patient while I worked for many hours over the past four years writing this book.

I thank the leadership at Perkins+Will and many of my co-workers for supporting me throughout this effort. The firm continues to support grass-roots research and there was always someone willing to help me on a daily basis. There were never any internal hurdles, and that allows for a successful work environment and encourages innovation for all of us. Mimi Day did a great job with the graphic design layout. Patty Gregory managed the entire process over the last four years and Meg Thorton assisted with the editing.

I also credit Michel Diot, who was a great client and friend for almost 20 years. I had the wonderful opportunity to work with Michel in the early 1990s designing a research center for Glaxo near Versaille, France. About half way through writing this book I asked Michel to give me his comments. He gave me enthusiastic support, and strongly encouraged me to continue. This is the way Michel always worked—in great support of each person he met. Unfortunately Michel passed away in the fall of 2008—he is missed but not forgotten.

I asked several top researchers, many of whom I am currently working with on projects, to share their expertise. I also asked them how they thought their areas of research would develop over the next 10 years. I thank the following people for their contribution to the book and for adding another layer of information that is truly appreciated:

Chapter 01 Walter L. Bradley—PhD, PE, and distinguished professor of mechanical engineering at Baylor University, Waco, Texas.

Chapter 03 Tim Maher, researcher at OMRF helped articulate the drug discovery process.

Chapter 04 *Energy research:* Otto Van Geet—PE, National Renewable Energy Laboratory in Boulder, Colorado. *Green chemistry:* Paul Anastas—assistant director of the Environmental Protection Agency.

Chapter 05 *Personalized medicine:* Dr. Leroy Hood—president of Systems Biology in Seattle, Washington. *Biobanks:* Dr. Mohammed Al Jumah—director for King Abdullah International Medical Research Center, Saudi Arabia.

Chapter 06 Karl Johnson—researcher and founding chief (retired) of the Special Pathogens Branch at the Centers for Disease Control and Prevention.

Chapter 07 *Neuroscience research:* Dr. Robert Finkelstein—National Institute of Neurological Disorders and Stroke at the National Institutes of Health. *Nanotechnology:* Dr. Zhong Lin Wang—director, COE, and distinguished professor at Georgia Institute of Technology, Atlanta, Georgia.

Chapter 08 *Translational research:* Dr. David S. Stephens—director of medical research at Emory University, Atlanta, Georgia. *Virtual reality:* Dr. Patrick S. Bordnick—director of the Virtual Reality Clinical Research Laboratory at the University of Houston, Houston, Texas.

Chapter 09 Jeannette Yen—professor at the School of Biology, and director of the Center for Biologically Inspired Design at the Georgia Institute of Technology, Atlanta, Georgia.

And finally, I would also like to acknowledge and thank all the researchers who are working tirelessly to improve our lives and help extend life expectancy for all people.

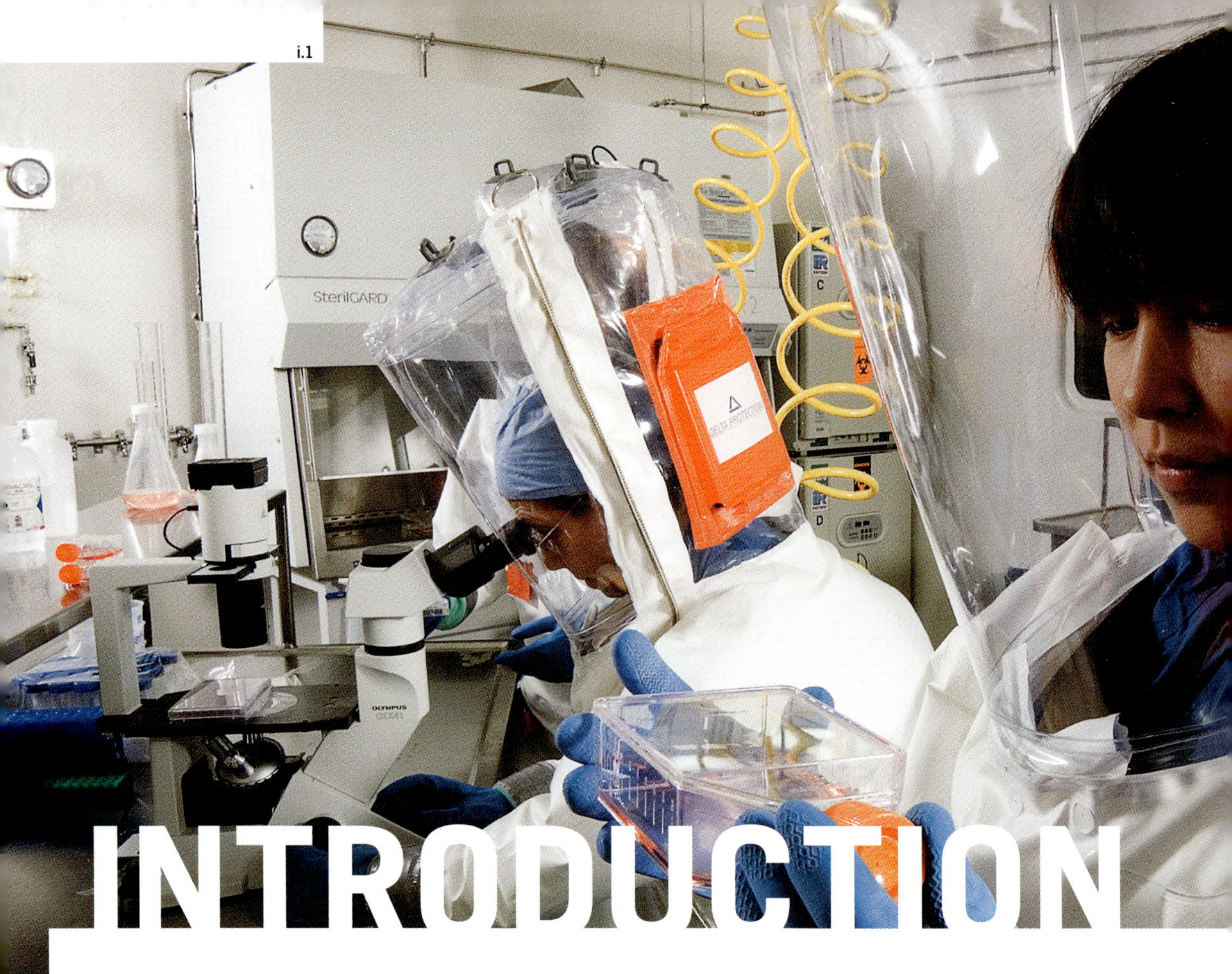

INTRODUCTION

Scientific discoveries are transforming our lives and the economies of nations worldwide. We live in an amazing time. In exploring the dynamic changes taking place in research—and the monumental impact on our lives—I hope to shed light upon the vast research opportunities at hand and encourage you to join with others in support of research, whether through financing, political decision-making, education, laboratory construction, or through conducting the actual research. To maximize scientific potential for the greater good, leaders of private industry and academic institutions, politicians, philanthropists, and scientists must all work together.

Science and scientists are usually not well understood by the general public. **This book is intended to communicate to all people of all ages the benefits of science today and potential opportunities for tomorrow.** Communication between scientists and the general public should be improved to lead to a greater acceptance of the benefits of supporting continued scientific research.

With technological advances, scientific knowledge has grown exponentially in recent years, resulting in a host of "new sciences." Genomics, proteomics, bioinformatics, nanotechnology, robotics, and a growing list of scientific concentrations show tremendous potential to radically alter medicine, agriculture, manufacturing, education, and more. If there are two questions that many people would like to know the answer to early in their life they would be, "How will I die and is there anything I can do to extend my life?" Research is now beginning to provide some answers to these questions; however, to allow most people to be tested and treated the answers need to be affordable.

Much of the research being conducted today is focused on eliminating or minimizing diseases in order to extend and improve our lives as well as championing environmental issues. The key for many investors and research institutions is to determine which research studies are the most important. In the USA the baby boom generation has money to donate to research; governments in other countries are generating strategies to develop research within their country to improve their citizens' quality of life, and pharmaceutical and biotech companies are trying to develop more successful "hits," or research discoveries to create profits for shareholders.

Choosing and then developing the right areas of research and development are extremely important decisions. We do not necessarily need to spend more money on research but instead we need to be more effective in the way research is done through smarter collaboration, by prioritizing research programs, and by planning more strategically from a global perspective. The "science" of science will need to be improved and be more cost effective. Many of the ideas and information throughout this book will help in the decision-making process.

Science holds the promise of significantly improving the quality and duration of all of our lives, especially for the world's poorest and most underserved populations. Medicine is being revolutionized. Many leading scientists believe we're on the cusp of treating, and preventing, some of the world's most common diseases, including heart disease, diabetes, Parkinson's disease and Alzheimer's disease. Personalized medicine is 10 to 15 years from being implemented to a large percentage of the population. Science is no longer the stronghold of a few privileged countries. Nations worldwide have awakened to the fact that economic growth is now fueled by knowledge and ideas, and that realization is transforming global economics. This book offers a snapshot of international research activities and priorities. The world's developed countries are aggressively funding research and are hungry for work. Many developing countries are focused on becoming players in the global economy.

While on a business trip to Riyadh in the summer of 2009, I read the newspaper headline "AIDS Breakthrough." Headlines announcing these kinds of research discoveries are almost a daily occurrence and have an impact on all people around the world.

In many ways, the scientific process has been turned on its head. Forget the image of the lone scientist toiling away in isolation. Today, science is collaborative on a global scale. Alliances are multinational and strategic, yet competition is fierce and speed is all-important. The

i.1 / BSL-4 HIGH-CONTAINMENT LABORATORY IN GALVESTON, TEXAS, ILLUSTRATING THE SIGNIFICANT CHANGE IN HOW RESEARCH IS DONE IN THE 21ST CENTURY.

i.2 / THOMAS EDISON'S LABORATORY AT THE TURN OF THE 20TH CENTURY, LOCATED IN FT. MYERS, FLORIDA.

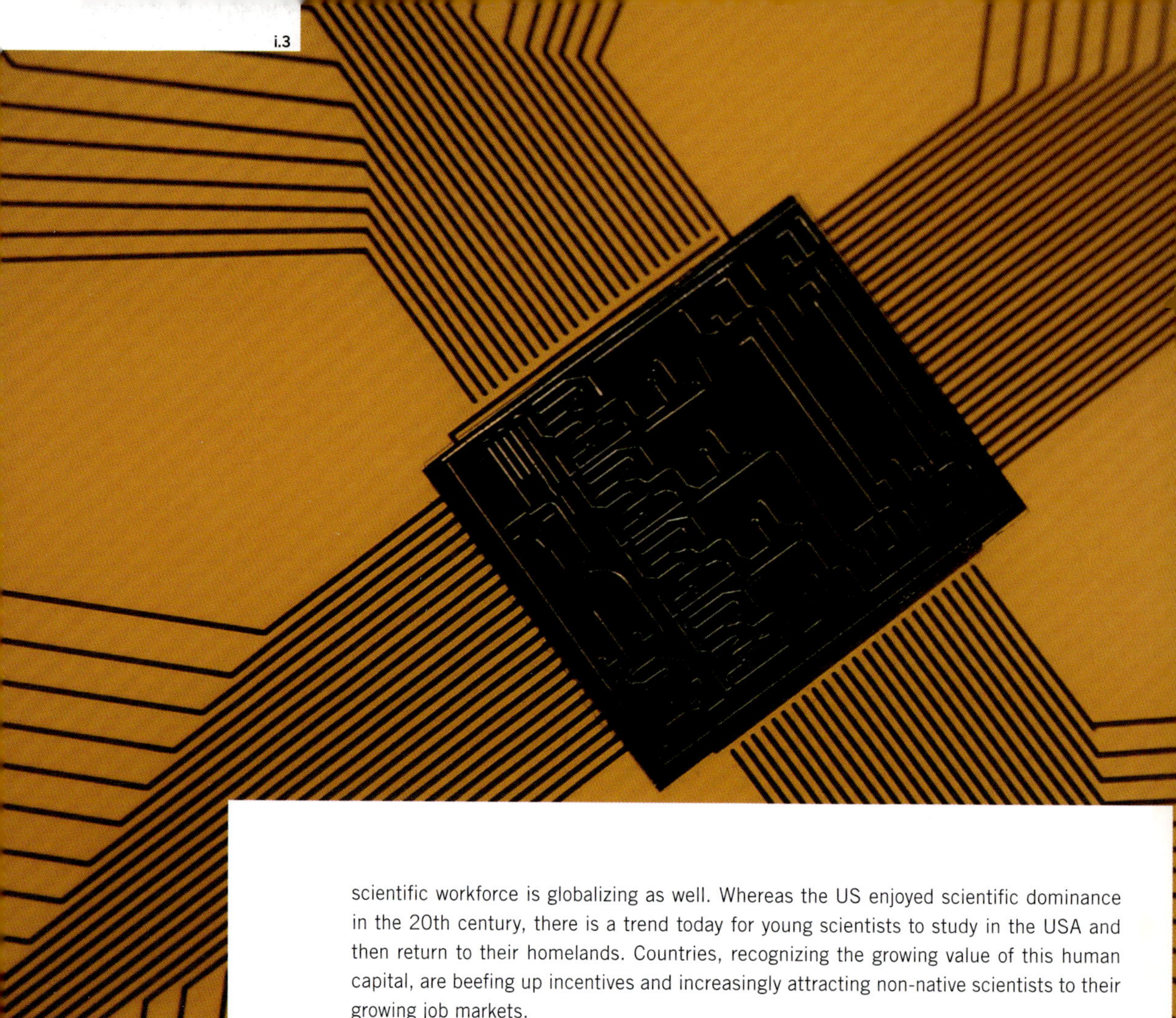

scientific workforce is globalizing as well. Whereas the US enjoyed scientific dominance in the 20th century, there is a trend today for young scientists to study in the USA and then return to their homelands. Countries, recognizing the growing value of this human capital, are beefing up incentives and increasingly attracting non-native scientists to their growing job markets.

International collaboration is credited for the wildly successful Human Genome Project, which mapped 99.9 percent of the DNA code and gave rise to the exciting new field of proteomics. A few research groups from various countries worked together to complete—ahead of schedule—the sequencing of the DNA model. That rich databank of scientific information is now on the Internet for anyone to access.

As stated by Stanford economist Paul Romer, the raw material of economic growth in coming years will not be physical labor or natural resources, but ideas that are hatched and developed in universities, research centers and other knowledge-based settings. Many companies and institutions are focused on research for the global good to more integrally support the economic, social, and cultural health of our world.

Sharing and leveraging brainpower to improve the quality of all our lives will hopefully improve relationships between countries. Scientific knowledge is a highly respected and an increasingly shared global commodity. One day, perhaps, the nations of the world will value the knowledge acquired through research more than any other resource. It is my hope that this book will help accelerate research discoveries, and it is my belief that effective collaboration and thoughtful partnerships provide the best opportunity to do so. These partnerships should be global between governments, academic institutions, and private industry. The research discussions in this book primarily focus on the life and

i.4

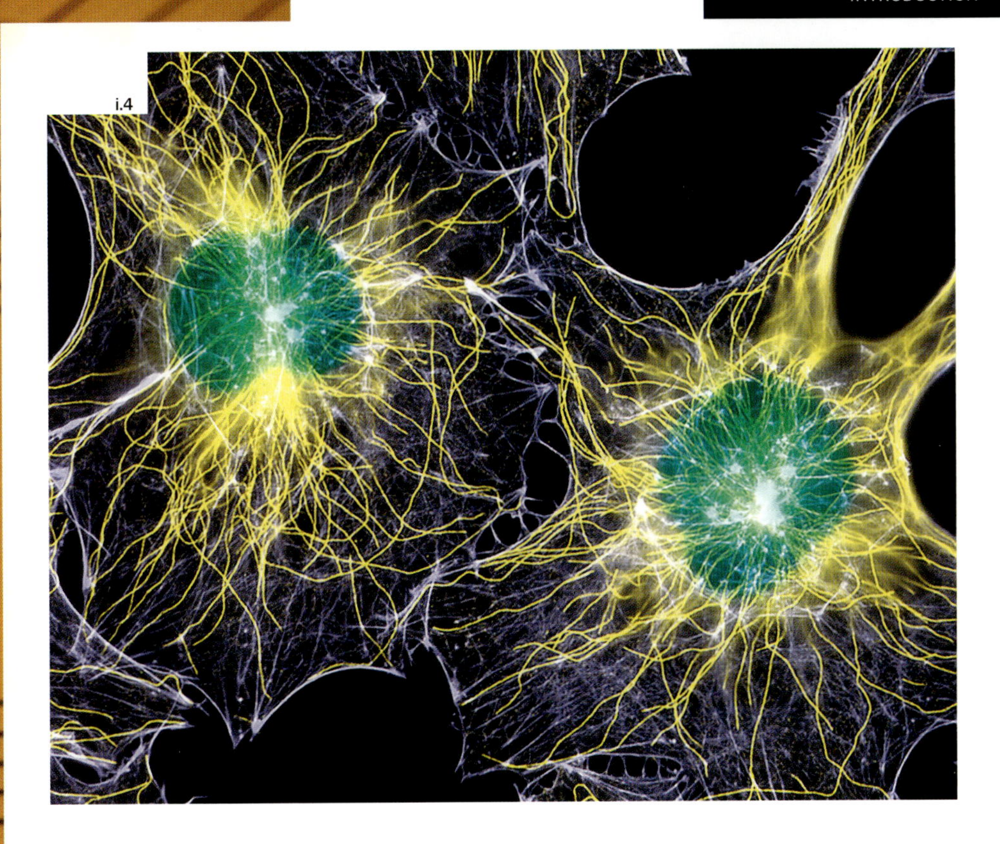

environmental sciences. Areas such as space research and historical studies, for instance, are not discussed but contribute to the larger research community.

Chapters 1–3 cover key research trends and the impact of research on the global economy. Developing countries increasingly view research and development as the key to accessing the fruits of the global economy and improving the lives of their citizens. At the start of this century, the National Science Foundation reported that, globally, approximately 50 percent of research was financed by the USA and 20 percent was financed by Japan, with the remaining 30 percent financed by all other countries. Now aided by technology, advances in science, and a heightened awareness of research and collaborative opportunities, the global research landscape is changing as nations commit their most important resources to developing their scientific capabilities.

A true global research partnership, however, requires the participation of all able nations. Furthermore, wealthier countries are obliged to help the world's poorest countries establish research and development infrastructures designed to become self-sustaining. Without a basic business model to support growth, poor countries will continue to struggle and the potential for violent unrest will escalate. Research in poor countries should address basic health and quality-of-life issues, as well as create local jobs and, most importantly, sustain hope.

Chapters 4–7 examine sustainable research, the success of collaborative international research, politically motivated sciences, and some of the breakthroughs expected from "new" sciences. These new sciences have a history of only a few decades, and their real potential is just now coming to light. Expect many important discoveries over the next 10 years. Also expect more new sciences to emerge as technology improves, as discoveries are

i.3 / THE 21ST CENTURY WILL CONTINUE TO SEE RAPID CHANGE. THIS MICROELECTROMECHANICAL SYSTEMS CHIP, SOMETIMES CALLED "LAB ON A CHIP," MAY SOME DAY HELP DOCTORS RAPIDLY DIAGNOSE A WIDE RANGE OF CONDITIONS IN THE CLINIC.

i.4 / IMAGING EQUIPMENT DEVELOPED OVER THE PAST QUARTER OF A CENTURY IS ABLE TO VISUALLY REPRESENT RESEARCH AREAS OF STUDY IN GREAT DETAIL. ACTIN (PURPLE), MICROTUBULES (YELLOW), AND NUCLEI (GREEN) ARE LABELED IN THESE CELLS BY IMMUNOFLUORESCENCE. THIS IMAGE WON FIRST PLACE IN THE NIKON 2003 SMALL WORLD PHOTO COMPETITION.

made and as investments in research increase. Research funding by country is discussed, as well as investments in new types of research. We also look at spending by nations, private corporations, and academic institutions. An examination of investments and future projections may help shed light on opportunities to collaborate and thus enhance outcomes.

Chapter 4 covers research related to sustainability, a key focus worldwide due to global warming, high energy costs, opportunities for new discoveries, and the world's increased use of energy, especially with the rapid growth of developing countries. US President Barack Obama's stimulus package has changed the US focus more toward energy efficient solutions and development of renewable energies to support the national grid. This is a sharp contrast to former President George W. Bush's focus on biodefense. Researchers from all over the world are trying to create more effective solar panels, fuel cells, and many other technologies that are better for the economy and the environment. Another primary focus is to reduce waste, become more efficient, and save money. New discoveries in these areas will be key drivers in the global economy for the next 10 to 12 years. In addition to offering environmental benefits, "green collar jobs" are expected to create employment opportunities in developing countries. Green research discoveries will be faster to market than discoveries focused on human health primarily because the review process will be easier than that required to gain approval for new drugs.

A combination of need, the ability to be quicker to market, and global opportunities will encourage a strong

FOCUS ON GREEN RESEARCH

in the next decade.

In 2009, the US, Japan, China, and the European Community focused on energy technology and clean energy programs as key components of their stimulus packages as a way to help address the global financial crisis.

Genomics, proteomics, and bioinformatics are presented in Chapter 5. Genomics is at the forefront of much research since the completion of the majority of the Human Genome Project in 2003. The research community now has vast amounts of data that brings an important new dimension to biological and biomedical research. Most of the data are now in the public domain. Personalized medicine will soon be available to many in wealthier countries. One challenge is to provide affordable solutions.

Proteomics, the next step after genomics, is the large-scale study of proteins, particularly their structures and functions, and how they interact. While the genome is an almost constant entity in a single organism, the more complicated proteome differs from cell to cell and is constantly changing through its biochemical interactions with the genome and the environment. The study of proteomics requires at least 10 times more data than the genomics database.

Bioinformatics involves the collection, storage, and analysis of biological information. Biological problems are solved using techniques from applied math, computer science, statistics, physics, and chemistry, which allows information to be understood and organized on a large scale. Most of the new sciences require powerful technology to analyze vast amounts of data and to maximize the volume and speed of the research within the dynamic and competitive international economy. This is a new challenge in research today—determining how to effectively create data, study it, and then use the information to accelerate the research process.

Chapter 6 looks at stem-cell and biocontainment research, both somewhat controversial and politically charged. Stem cell research has resulted in amazing discoveries with tremendous medical potential. Stem cells are the parent cells for all tissues and organs, and they exist to maintain and repair blood, bone marrow, skin, muscle, and organs such as the brain and liver. Research on embryonic stem cells raises ethical questions because extracting the stem cell destroys the embryo, and thus destroys the potential for that life even though it may extend the lives of others. The debate over stem-cell research has been most pronounced in the US. The Obama administration supports many types of stem-cell research.

Since the attack on New York's World Trade Center on September 11, 2001, funding for biocontainment research and laboratories has skyrocketed. So far in the 21st century,

i.5 / AUTOMATION AND HIGH-THROUGHPUT LABORATORIES WILL BE VERY COMMON IN THE 21ST CENTURY TO ALLOW FOR MORE RESEARCH TO OCCUR AT A FASTER PACE.

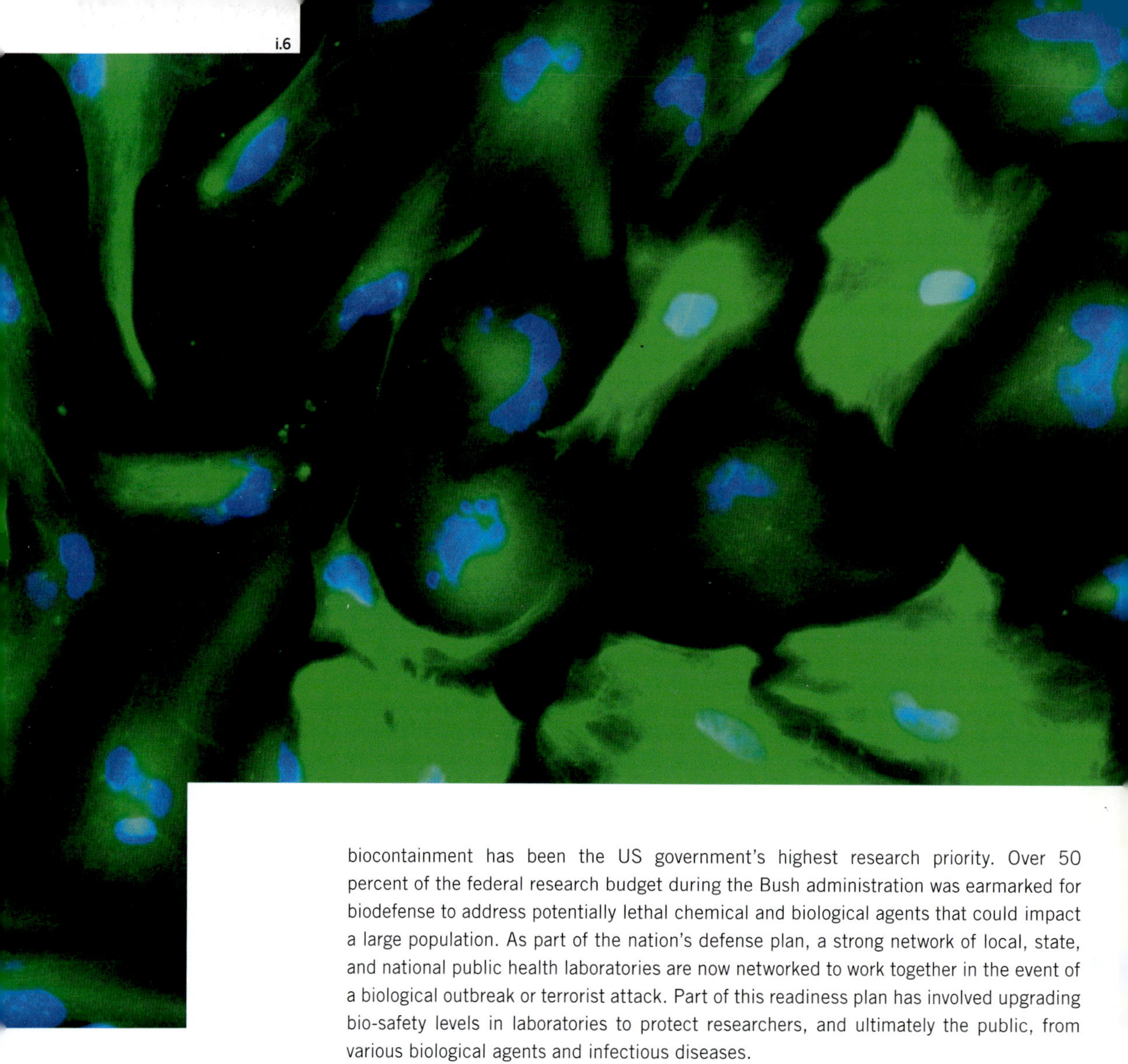
i.6

biocontainment has been the US government's highest research priority. Over 50 percent of the federal research budget during the Bush administration was earmarked for biodefense to address potentially lethal chemical and biological agents that could impact a large population. As part of the nation's defense plan, a strong network of local, state, and national public health laboratories are now networked to work together in the event of a biological outbreak or terrorist attack. Part of this readiness plan has involved upgrading bio-safety levels in laboratories to protect researchers, and ultimately the public, from various biological agents and infectious diseases.

Chapter 7 focuses on nanotechnology and neuroscience. Nanotechnology is engineering on the molecular scale, that is, using molecular-scale structures as the components of future systems that are designed and integrated on that same scale. This involves placing molecules and other molecular-scale structures, such as nanowires and quantum dots, where required in enormous numbers. Nanotechnology will make products lighter, stronger, smarter, cheaper, cleaner, and more precise. Nanotechnology discoveries can have a significant impact on the global energy crisis by creating more efficient materials and solutions that will require less energy and less waste.

Neuroscience is the study of the nervous system, including the brain, the spinal cord, and networks of sensory nerve cells, or neurons, throughout the body. Neuroscientists use tools ranging from high-resolution imaging equipment and other computers to special dyes to examine molecules, nerve cells, networks, brain systems, and behavior. From these studies, they learn how the nervous system develops and functions normally, and what goes wrong in neurological disorders. There are approximately 600 neurological disorders that impact about one-sixth of the US population each year.

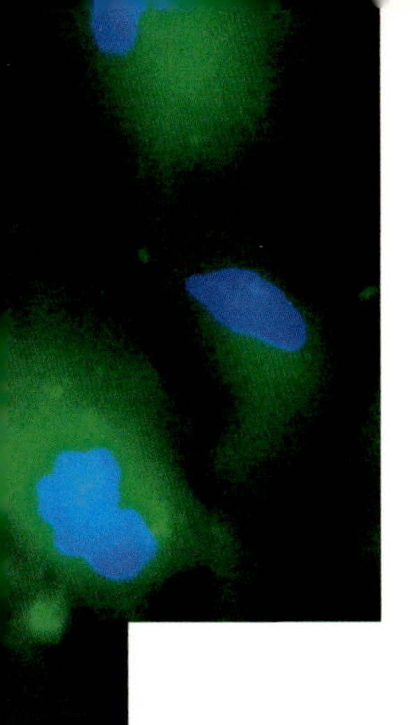

There is a need with the science industry to
REEVALUATE AND REINVENT
the way research is conducted.

Chapter 8 is about improving the science of science. The chapter highlights the dramatic changes in research processes to support, model, and accelerate research. This has been brought about largely due to the development of computer technology. Today's high-throughput labs can process tens of thousands of samples daily to facilitate quick responses to understanding a virus or helping to create a vaccine. These flexible labs can be programmed to focus on one study or allow rapid testing of multiple samples. There is the need with the science industry to reevaluate and reinvent the way research is conducted. There has been too much waste, time, and cost to get to a discovery or to stop a study. There is the opportunity now for better collaboration and equipment to work with. Improving the "science of science" is required today to just compete.

Interdisciplinary research is revolutionizing science as researchers now routinely collaborate with the well-founded expectation that new ideas will develop from multiple scientific fields. Behavioral sciences involve three-dimensional virtual models of spaces that simulate a specific activity to better understand how a patient reacts to a specific issue or condition. Highly sophisticated imaging equipment is also helping to provide answers and opportunities. Translational research is the bridge from discovery to delivery. In medicine it has a clinical goal or target, while in other fields there are more practical applications such as plant genome, improved crops, and biofuels. The process of translational research is often thought of as a linear process, starting with a discovery in the lab and then moving straight to the patient or application, but under optimal conditions the process goes back and forth until the appropriate solution is determined. So in medicine, for example, the project begins in the clinic, a pathway identifies a problem in patients, takes that observation back to the lab, designs a solution, and then works through clinical trials and regulatory approval to deliver the final solution to patients.

i.6 / STEM CELLS ARE THE PARENT CELLS FOR ALL TISSUES AND ORGANS.

i.7

Chapter 9 focuses on education—the international foundation of success for individuals as well as for nations. Recent models for education include hands-on research, interactive classrooms, and the use of the latest technologies. Universities are emerging as regional leaders while creating national and international partnerships with governments, private industry, and other universities. The international university model is blossoming and flourishing. Many developing countries, especially those with large oil reserves, are building several new universities and infrastructure projects.

Chapter 10 explores new ideas that are suggested to address many of the opportunities facing research institutions. The second half of the chapter discusses research parks around the world. Today's state-of-the-art research parks are developing strong brand images that are marketable in the global economy as well as effective in promoting new research models for the 21st century. The latest planning trends are discussed with photographs of recently completed projects.

Throughout the book there are informative quotes from leading international researchers and relevant data, most from within the past three years. As an architect, I have focused on presenting strong images to support the best research to reinforce the amazing studies that are happening, and discoveries that have been made. I believe you will see information presented in a creative way.

Beyond providing information, this book seeks to inspire readers to support research that can improve the lives of people worldwide. Please share this information with others and email me any thoughts or ideas that you would like to share to help me create a second edition with updated information. I can be reached at dan.watch@perkinswill.com.

Thank you!
Dan Watch

i.7 / RESEARCH LABORATORIES CONTINUE TO HAVE MORE AUTOMATED EQUIPMENT TO HELP STUDY MORE OPPORTUNITIES FASTER AND MORE EFFECTIVELY.

i.8 / GECKOS CAN RUN UP SMOOTH SURFACES, WHICH CAN HELP LEAD TO DISCOVERIES IN ADHESIVES AND ROBOTS. NATURE IS PROVIDING US WITH WONDERFUL EDUCATIONAL OPPORTUNITIES AND RESEARCH DISCOVERIES IN THE FIELD OF BIOMIMICRY—DISCUSSED IN CHAPTER 9.

01. SCIENCE IS GOLDEN

How Research Impacts on the Global Economy /

Scientific research is having a tremendous impact on the economy of nations and the world at large. It is also extending and improving our lives. Because of this, knowledge and innovation are now the most important assets of any person, business or country. To optimize opportunities for research discoveries, we must understand global influences. This chapter looks at macro issues impacting research around the world.

The 21st Century Brings Dramatic Changes in Science /

At the turn of the millennium, I wrote a book entitled *Research Laboratories* (John Willey & Sons, 2001) detailing the architectural and engineering aspects of constructing state-of-the-art laboratory facilities. That book was released in early October 2001, less than a month after September 11. Since then, much has happened in the world of science:

The initial computer modeling for the **Human Genome Project** was completed in April 2003.

On the heels of genomics, **proteomics** is already facilitating medical breakthroughs, in part owing to the amazing amounts of data that can be analyzed on new supercomputers. Bioinformatics, a relatively new industry, supports the development and interpretation of these data.

Stem cells, which we now know can repair damaged tissues, show great promise in treating and preventing diseases.

Nanotechnology, the new blue sky research, presents vast opportunities in everything from energy efficiency to medicine to clothing.

Neuroscience, the study of the nervous system, has made it possible to understand, in exquisite detail, the complex processes that produce intellectual behavior, cognition, and emotions. The field offers great hope in addressing brain disorders, which affect about one-sixth of the world's population.

In the wake of the terrorist attacks on the World Trade Center and the Pentagon in 2001, biocontainment laboratories have been constructed throughout the US.

Sustainability has emerged as a key global issue.

I started writing this book in the summer of 2007. Since then there have been three significant changes that made me go back, update, and research many areas of this book. Throughout the book these issues are discussed:

- The global economic downturn
- The global green movement
- The election of US President Barack Obama.

In March 2009, President Obama stated that "medical miracles do not happen simply by accident. They result from painstaking and costly research, from years of lonely trial and error, much of which never bears fruit, and from a government willing to support that work. From life-saving vaccines, to pioneering cancer treatments, to the sequencing of the human genome—that is the story of scientific progress in America."

Scientific Partnerships Benefit All Nations /

We live in a world of daunting challenges—limited natural resources, global warming, and in many areas extreme poverty. With the emergence of technologies and the new sciences, important discoveries occur almost daily.

For the first time in history, significant investments are being made in global well-being. Through global initiatives, medical access and other basic necessities are being provided to some of the world's poorest people. The foundation of such activity is thoughtful partnerships among people, companies, universities, and nations. The sharing of information on the Internet in real time has given rise to these dynamic partnerships and allows them to evolve quickly. Research is now a global phenomenon and a global industry, and smart, effective partnerships are the key to accelerating discoveries. These partnerships will benefit some people financially, but, ultimately, research discoveries should improve everyone's quality of life.

Bill Gates, former CEO of Microsoft, is encouraging "creative capitalism," which means global corporations are the distinguishing feature of modern capitalism and as such they should integrate do good policies into the way they conduct business.

01.1 / A GROUP OF GHANAIAN STUDENTS LEARNING HOW TO PROTECT THEMSELVES FROM GUINEA WORM DISEASE (GWD) THROUGH THE INSTRUCTION OF A RED CROSS VOLUNTEER. THE VOLUNTEER USES A WALL MURAL PAINTED ON THE SIDE OF A GUINEA WORM-CASE-CONTAINMENT CENTER THAT DESCRIBES GWD, ITS SYMPTOMS, AND THE METHODS REQUIRED TO AVOID INFECTION. IMAGE COURTESY OF THE CARTER CENTER AND THE GHANA GUINEA WORM ERADICATION PROGRAM.

Gates explained in a January 2008 presentation in Davos, Switzerland, "that the world is getting better, but it's not getting better fast enough, and it's not getting better for everyone ... in particular the billion people who live on less than a dollar a day." He goes on to say, "we need a system that draws in innovators and businesses in a far better way than we do today. Such a system would have a twin mission: making profits and also improving the lives for those who don't fully benefit from market forces."

Gates calls this new system **"creative capitalism—an approach where governments, businesses, and nonprofits work together** to stretch the reach of market forces so that more people can profit, or gain recognition, doing work that eases the world's inequities." Put another way, creative capitalism is interested in the fortunes of others and ties it in with the interest we

01.2

have in our own fortunes, in ways that help advance all concerned. For instance, people who are experts on the needs of developing countries would meet several times a year with scientists from software or drug companies and help try to find poor-world applications for their best ideas. Another approach to creative capitalism includes a direct role for governments. Some of the highest leverage work that governments can do is to set policy and disburse funds in ways that create market incentives for business activity that improves the lives of the poor.

Tiered pricing is an incentive that drug companies should consider adopting. This means a drug company would take a valuable patent and charge a full monopoly price in the developed world, but let manufacturers in developing countries produce the product for less than a dollar a dose. This is an idea that has been suggested by Warren Buffett.

Under a law signed by former President George W. Bush in 2007, any drug company that develops a new treatment for a neglected disease like malaria or tuberculosis (TB) can get priority review from the Food and Drug Administration for another one of its products. So, for example, if a company develops a new drug for malaria, its profitable cholesterol-lowering drug could go on the market a year earlier. This priority review could be worth hundreds of millions of dollars. There is a growing understanding around the world that when change is driven by market-based incentives, a sustainable plan for change is established.

Gates also recommends to "create measures of what companies are doing to use their power and intelligence to serve a wider circle of people. This kind of information is an important element of creative capitalism. It can turn good works into recognition, and ensure that recognition brings market-based rewards to businesses that do the most work to serve the most people ... If we can spend the early decades of the 21st century finding approaches that meet the needs of the poor in ways that generate profits and recognition for business, we will have found a sustainable way to reduce poverty" and increase life expectancy for a global population.

01.2 / POVERTY CAN BE REDUCED BY USING RENEWABLE RESOURCES LIKE COCONUTS THAT CAN BE GROWN BY LOW-INCOME FARMERS IN DEVELOPING COUNTRIES.

01.3 & 01.4 / CHILDREN ON A COCONUT FARM.

While I was starting to design the new Baylor Research and Innovation Center in the fall of 2009 their faculty shared a wonderful story of the students and faculty doing work around the world, especially in developing areas. "We set out to discover the technology that would allow us to start a triple-bottom-line company based around coconuts called Whole Tree Inc, one that would make a profit, make a difference in the lives of people most in need, and be good stewards of our planet" explained lead researcher Dr. Walter Bradley.

One day soon the car you buy may be made partly out of coconuts, with a farmer thousands of miles away reaping great benefit. If a plan developed by engineering students and their professor at Baylor University lives up to expectations, that farmer will triple his annual income from about US$500 to US$1500.

It all started a few years ago when engineering professor Walter Bradley set out on a very specific search. That search eventually led him to the Philippines and, of course, to the coconut. The coconut palm is ubiquitous throughout the tropics, and while many uses have been found for the nut the tree produces, as well as the husks that encase the nut's hard shell, they are of such meager commercial value these days that a typical coconut farmer in the Philippines, the world leader in coconut production, earns only about 10 cents for every coconut.

The research team at Baylor University have identified a variety of low-cost products that can be manufactured from coconuts in poor coastal regions and have now developed a way to use coconut husks in automotive interiors. The Baylor researchers are developing a

technology to use coconut fiber as a replacement for synthetic polyester fibers in compression-molded composites. Specifically, their goal is to use the coconut fibers to make trunk liners, floorboards, and interior door covers on cars, marking the first time coconut fibers have been used in these applications.

Since coconuts are an abundant, renewable resource in all countries near the equator, Baylor's team is working to create multiple products that could be manufactured from coconuts in those regions using simple and inexpensive technology. With an estimated 11 million coconut farmers in the world making an average annual income of $500, the Baylor researchers hope to triple the farmers' annual income by increasing the market price for each coconut to 30 cents.

"What we hope to do is create a viable market for the poor coconut farmer," said Dr. Walter Bradley, distinguished professor of engineering at Baylor, who is leading the project. "Our goal is to create millions of pounds of demand at a much better price."

The research team focused on creating innovative technologies that add value to the local community. The Baylor researchers said the mechanical properties of coconut fibers are just as good, if not better, than synthetic and polyester fibers when using them in automotive parts. Dr. Bradley said the coconut fibers are less expensive than other fibers and better for the environment because the coconut husks would have otherwise been thrown away. Coconuts also do not burn very well or give off toxic fumes, which is crucial in passing tests required for actual application in commercial automotive parts.

Dr. Bradley said they are working closely with Hobbs Bonded Fibers, a Waco, Texas-based fiber-processing company that is a supplier of unwoven-fiber mats to four major automotive companies.

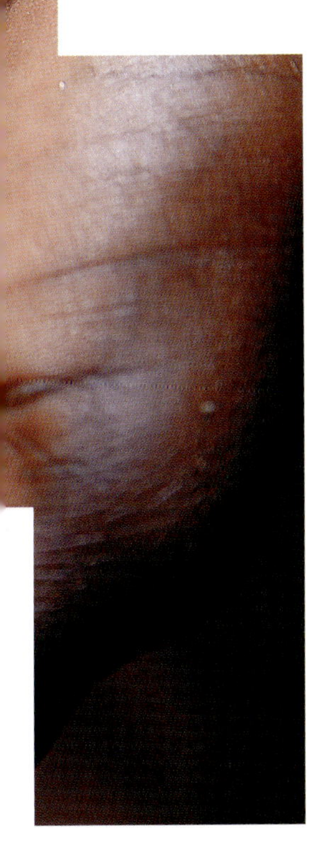

Fast Company and Monitor Group's second annual Social Capitalist Awards identified 25 organizations that are using creativity, business smarts, and hard work to invent a brighter future. Consider the work of Endeavor Global, a New York-based nonprofit organization that seeds economic growth in developing countries by supporting the work of large-scale entrepreneurs. In 2002, 97 companies funded by Endeavor generated $332 million in revenue and created 8,562 jobs in Latin America. Or take Social Venture Partners (SVP), which has invented a powerful model of philanthropy that now operates in 23 cities, where investors pool $5,500 apiece along with their time to help local nonprofits. SVP's elegant, virtuous-circle model produces smarter, more engaged donors, and stronger nonprofit organizations.

Returning Social Capitalist Awards' winner New Leaders for New Schools (NLNS) improves education in inner cities one school at a time, by recruiting and training principals for low-income schools. NLNS schools already have shown gains in reading and math scores. More than 20 cities are creating their own principal-training programs with elements of the NLNS model. Such results signal a new era of enlightenment in the social sector. Together, these organizations form a distribution network for social innovation—the beginnings of a system for reinventing systems.

Today, the research community, and the business concerns underpinning much of it, understands that collaboration maximizes success. It's not hard to grasp; through technology, information is now moving nonstop to all parts of the world. "Eighty percent of the scientists, engineers, and doctors who ever lived are alive today—and (they are) exchanging ideas in real time on the Internet," wrote Jay J. Jamrog, executive director of the Human Resource Institute, in an article titled "The Changing Nature of People".[1]

Increased collaboration among the world's scientists and the resulting development of ideas and innovations will be the greatest force in spreading prosperity globally.

01.5 / THE PROGRAM FOR AFRICA'S SEED SYSTEMS (PASS) IS AN INITIATIVE LAUNCHED WITH SUPPORT FROM THE BILL AND MELINDA GATES FOUNDATION THAT SEEKS TO PROVIDE QUALITY SEEDS TO RURAL AFRICAN FARMERS.

In *The World Is Flat,* Thomas Freidman reinforced the economic impact of shared knowledge when he wrote, "Flattening of the world means that we are now connecting all the knowledge centers on the planet together into a single global network, which—if politics and terrorism do not get in the way—could usher in an amazing era of prosperity and innovation."[2] Many times throughout history research has uncovered solutions for diseases and other problems only to see political and social problems prevent those solutions from benefiting the wider population.

The Rise and Ebb of the USA's Research Dominance /

The international science landscape is rapidly changing. In the past a country's economic and scientific global standing shifted gradually, usually over centuries. But in the age of globalization, a nation's standing can change within a single generation.

The US has dominated the world in science since around the mid-1900s. Historically speaking, however, the US's research stronghold is a relatively new phenomenon. From 500 AD to 1500 AD, the Middle East led the world in science. Following the rise of Islam in the 7th century, science and technology flourished in the Islamic world. Muslim leaders encouraged exploration in mathematics, astronomy, medicine, pharmacology, optics, chemistry, botany, philosophy, and physics. Much of the knowledge developed by the Muslims was transmitted to the Europeans, who emerged from the Dark Ages into the Renaissance. Europe then dominated science for some 500 years. The astronomical discoveries of Nicolaus Copernicus, Tycho Brahe, Johannes Kepler, Sir Isaac Newton, and Galileo are the basis of modern physics.

The Nobel Prize in Physiology or Medicine is awarded once a year by the Swedish Karolinska Institute. It is one of five Nobel Prizes established by the will of Alfred Nobel in 1895. The award is regarded as the most prestigious that a scientist can receive. The US has received more awards than any other country and more than the next five countries combined. Japan and China have won the two awards in Asia, but over the next 20 years I predict a significant rise in the number of awards won by Asian countries.

Most Nobel Laureates and other innovators do their most productive and creative work between the ages of 30 to 45. In the 20th century, the migration of the world's scientists to the US was dramatic. The first wave came when the Nazis drove a generation of world-class scientists out of Europe. Among them were Albert Einstein, Edward Teller, Enrico Fermi and Niels Bohr, as well as thousands of well-

Nobel Prizes in Chemistry/Physics*

1.	USA	92
2.	UK	41
3.	GERMANY	34
4.	FRANCE	16
5.	SWITZERLAND	10
6.	SWEDEN	9
7.	JAPAN	9
8.	RUSSIA	7
9.	NETHERLANDS	7
10.	CANADA	6

*1901–2008

01.6 / TODAY'S LAB HAS A WIDE RANGE OF TECHNOLOGY FROM STATE-OF-THE-ART FUME HOODS AND BIO-SAFETY CABINETS TO CLEAN ROOMS TO SPECIALTY EQUIPMENT. HERE AT NC STATE COLLEGE OF ENGINEERING, THE GRADUATE STUDENTS ARE WORKING WITH THE BEST LASER EQUIPMENT AVAILABLE.

trained scientists and future scholars. The next large influx came from scientists being driven out of Europe by communists. Among them was rocket scientist Wernher von Braun. Over the years, the US economy has been fueled by the contributions of throngs of immigrant scientists and engineers including: Andrew Carnegie (U.S. Steel) and Alexander Graham Bell (AT&T), both from Scotland; Andrew Grove (Intel) from Hungary; Vinod Khosla (Sun Microsystems) from India; and Sergey Brin (Google) from Russia.

In recent decades, war, political instability, and poverty in developing countries have driven many of the world's brightest students to the US for advanced education and work opportunities. Many remained, and by 2000 about half of the world's research was conducted in the US, much of it supported by US businesses. The US economy has benefited greatly from this international brain dump and the resulting groundswell of homegrown research discoveries.

Globalization, more than anything else, has reduced the numbers of extreme poor in India by 200 million and China by 300 million since 1990.

– *THE END OF POVERTY* BY JEFFREY SACHS, ECONOMIST AND PROFESSOR AT COLUMBIA UNIVERSITY

Today, however, we are seeing a marked shift in concentrations of scientific activity in the world. Driven by public health and economic growth initiatives, many countries are investing in education and research at unprecedented levels. Countries seeking competitive advantage are building science and technology infrastructures to attract foreign investment, import foreign talent, and develop regional and national expertise.

Technology has fueled standard of living increases in both prosperous and developing countries. This suggests that all of the world, including today's laggard regions, can benefit from technological advances. "Economic development is not a zero-sum game in which the winnings of some are inevitably mirrored by the losses of others," states economist Jeffrey Sachs in *The End of Poverty*. "This game is one that everyone can win."[3]

World Population Concerns /

After the Second World War, some 2.5 billion people inhabited the planet. Within the next 60 years the population more than doubled to over 6 billion. By 2050, the United Nations expects the world population to reach from 9 to 11 billion, depending upon fertility rates. This means an average of 50 million people are added to the world's population each year. Increased life expectancies are also fueling population growth. People aged 85 and over are now the fastest growing segment of many national populations. My wife's grandmother, Onie Ponder, reached her 111th birthday on September 3, 2009 and became the oldest person in Florida.

World population growth (2003–50)

(BILLIONS)	*2003*	*2050*
DEVELOPED COUNTRIES	1,203 (19%)	1,220 (14%)
LESS DEVELOPED	5,098 (81%)	7,699 (86%)
TOTAL	6,301	8,919
AFRICA	851 (13%)	1,803 (19%)
ASIA	3,823 (60%)	5,222 (59%)
LATIN AMERICA & CARIBBEAN	543 (09%)	768 (09%)
EUROPE	726 (12%)	632 (07%)
NORTH AMERICA	326 (05%)	448 (05%)
OCEANIA	32 (01%)	46 (01%)
TOTAL	6,301	8,919

01.7

The populations of China and India are each at least three times larger than the US, the third most populated country, and the population gap between less developed and more developed countries continues to widen. The annual rate of growth in developed countries is less than 0.3 percent, "while in the rest of the world the population is increasing almost six times as fast," according to Population Matters, a 2000 policy brief published by RAND, a nonprofit research institution.[4]

Europe's population is projected to decrease due to low birth rates. Meanwhile, Africa and Asia, already rife with poverty and/or infectious disease in many areas, will swell in population by 2050 if population growth is not minimized, the United Nations predicts. In some parts of the world, educational programs focused on population control are making headway as women learn about birth control and as cultures shift in greater support of women. If standards of living are to improve, these initiatives must expand and be actively supported by local governments.

Improvements in Life Expectancy /

As a leader in healthcare, it is surprising that the US ranks relatively low in life expectancy among wealthier nations. The US might offer the best healthcare if you need a complicated surgery, however, many other countries enjoy healthcare costs that are approximately half or less in comparison. Poor nations struggle, and some find ways to provide quality care on a modest budget while rich nations attempt to control excessive costs. The following graph identifies amounts countries spent in 2006 for healthcare based on the percentage of their gross domestic product (GDP).

The US spends the most money in the world on research and healthcare but is ranked 33rd in life expectancy. Most developed countries are ranked higher than the USA but spend less on healthcare. One key point of this book is that it's not how much money you spend, but how effective you are with the money you do spend. This may mean that

01.8 / HEALTHCARE SPENT AS A PERCENTAGE OF GDP (2006)

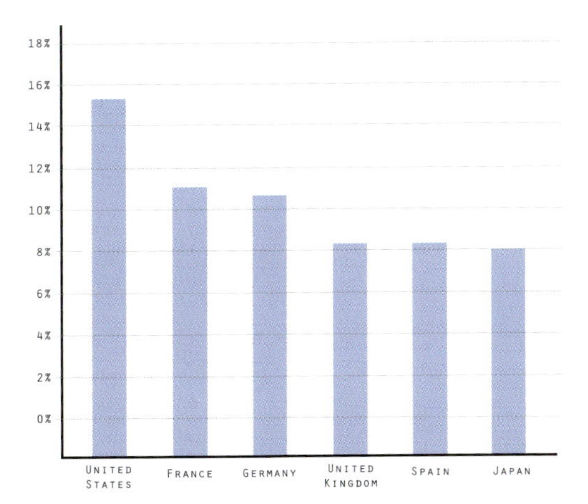

people in the US may be more effective helping people in developing countries if they are more accountable for their own health, whether through more exercise, fewer prescriptions drugs, or taking responsibility for some chronic diseases to name just a few. "Sixty-five percent of all healthcare costs in America could be eliminated by exercising more, losing weight, stopping smoking, and ending substance abuse," states G. Anderson in "Chronic Condition: Making the Case for Ongoing Care".[5]

It is also difficult to understand why it is acceptable globally for life expectancies in underdeveloped countries to be so low. Life expectancy in underdeveloped countries is about where developed countries were 150 years ago. Lessening the gap requires support of the global community. Japan has the longest average life expectancy at 83 years, followed closely by Andorra. The Centers for Disease Control reported in August 2009 that the average American born today could expect to live 78 years mainly due to the falling death rates in almost all leading causes of death. The average life expectancy in developing countries typically increases an average of three months each year.

01.7 / AN EXHIBIT AT THE MUSEUM OF SCIENCE AND INDUSTRY IN CHICAGO DISPLAYS A PHOTO OF ONIE PONDER (CENTER) WHO, AT THE AGE OF 111, BECAME THE OLDEST PERSON IN FLORIDA.

01.8 / HEALTHCARE SPENT AS A PERCENTAGE OF GDP: 1) USA 15.3; 2) FRANCE 11.1; 3) GERMANY 10.6; 4) UK 8.4; 5) SPAIN 8.3; 6) JAPAN 8.2.

01.9 /FOR AT LEAST 800 YEARS (1000–1800) THERE WAS A VERY SLOW INCREASE IN AVERAGE LIFE EXPECTANCY—APPROXIMATELY AN INCREASE OF ONE ADDITIONAL YEAR FOR EACH CENTURY.

01.9 / LIFE EXPECTANCY AT BIRTH (1000CE–2000CE)

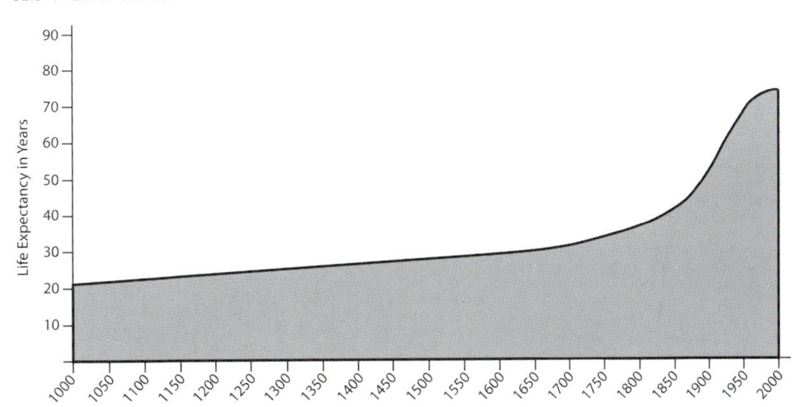

01.10

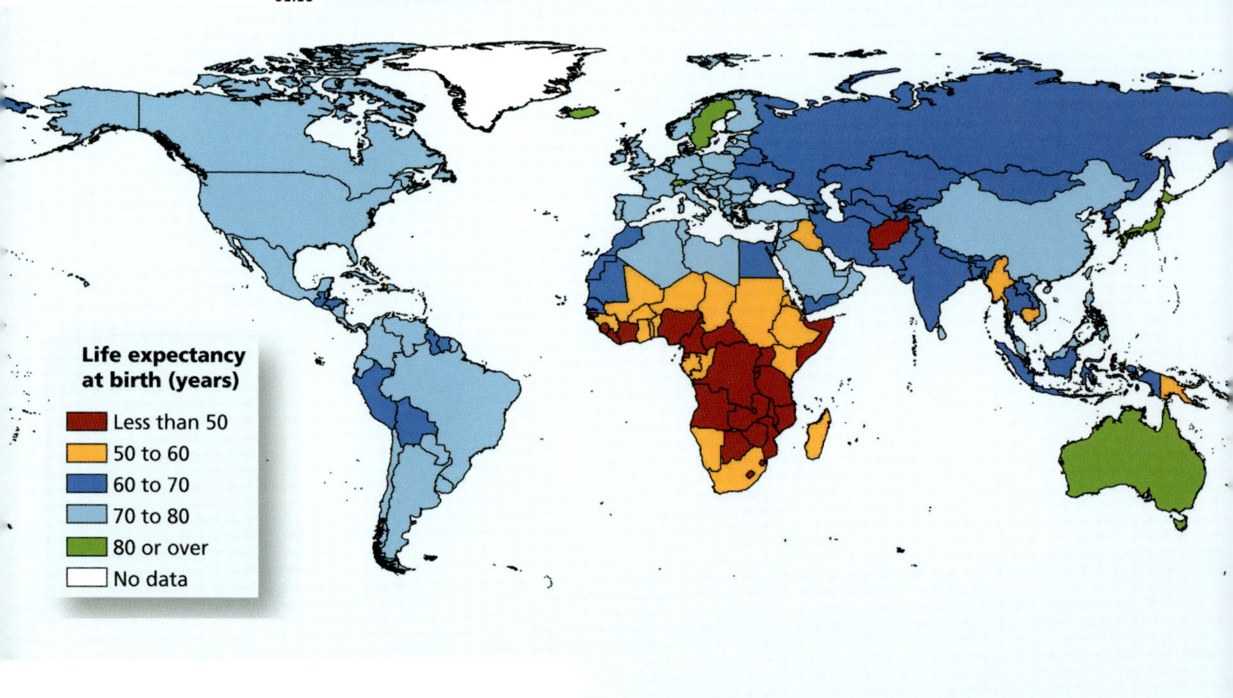

By the end of the 21st century, many people in developed countries are expected to live beyond 100 years. In the past 160 years, the life expectancy of people in developing countries has nearly doubled (see the following graph), with an average of three months added each year. However, AIDS has ravaged Africa, Asia and even Latin America by creating much lower life expectancy in 34 different countries (26 of them in Africa). African countries have the lowest life expectancies.

While I've been writing this book the life expectancy for developed countries has been relatively consistent averaging about three months of improvement each year. For developing countries, the recent data is quite different. Afghanistan now has the lowest life expectancy at 43.8 years, and this number has dropped from 47.7 in 2005 due to the ongoing war. There were 24 other countries just four years ago with lower life expectancy.

01.11

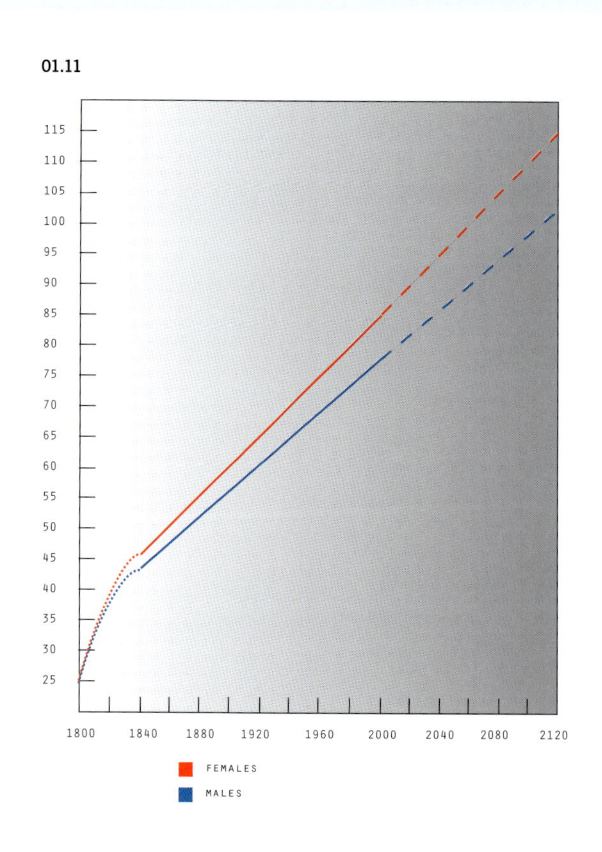

Highest Life Expectancy (2005–2010)

1.	JAPAN	82.7
2.	ANDORRA	82.5
3.	HONG KONG	82.2
4.	ITALY	82.1
5.	ICELAND	81.8
6.	SWITZERLAND	81.8
7.	AUSTRALIA	81.5
8.	FRANCE	81.2
9.	SPAIN	80.9
10.	SWEDEN	80.9
11.	CANADA	80.7
12.	ISRAEL	80.7
13.	MACAU	80.7
14.	NORWAY	80.6
15.	BERMUDA	80.4
16.	CAYMAN ISLANDS	80.4
17.	SINGAPORE	80.3
18.	NEW ZEALAND	80.2
19.	AUSTRIA	80.0
20.	NETHERLANDS	80.0
21.	GERMANY	79.9
22.	IRELAND	79.9
23.	BELGIUM	79.7
24.	CYPRUS	79.7
25.	MALTA	79.7
26.	FINLAND	79.6
27.	MARTINIQUE	79.6
28.	LUXEMBOURG	79.5
29.	FARCE ISLANDS	79.4
30.	SOUTH KOREA	79.4
31.	UK	79.4
32.	GREECE	79.2
33.	USA	79.2
34.	CHANNEL ISLANDS	79.1
35.	GUADALOUPE	79.1
36.	VIRGIN ISLANDS (US)	78.9
37.	COSTA RICA	78.8
38.	PORTUGAL	78.7
39.	PUERTO RICO	78.7
40.	CUBA	78.6
41.	CHILE	78.5
42.	SLOVENIA	78.4
43.	DENMARK	78.3
44.	TAIWAN	78.0
45.	KUWAIT	77.6
46.	UNITED ARAB EMIRATES	77.4
47.	BARBADOS	77.2
48.	BRUNEI	77.1

The other nine countries on the list have increased life expectancy from four to 15 years since 2006. Improved education, better healthcare, lower infant-mortality rates, birth control, and international support have made significant impact in many under-developed areas of the world. Ideally, if the developing countries continue to improve then by the end of the 21st century they may have the same life expectancy as developed countries. I hope I live to see the day that this comes true.

Lowest Life Expectancy (2005 & 2009)

		2005	2009
1.	AFGHANISTAN	47.7	43.8
2.	ZIMBABWE	37.3	44.1
3.	ZAMBIA	39.1	45.2
4.	LESOTHO	34.3	45.3
5.	SWAZILAND	29.9	45.8
6.	ANGOLA	41.9	46.8
7.	CENTRAL AFRICAN REPUBLIC	39.5	46.9
8.	SIERRA LEONE	41.9	47.4
9.	CONGO-KINSHASA	44.7	47.5
10.	GUINEA-BISSAU	41.5	47.6

01.10 / LIFE EXPECTANCY AT BIRTH, 2000–2005.

01.11 / LIFE EXPECTANCY AT BIRTH (1800–2100). OVER THE PAST 160 YEARS, LIFE EXPECTANCY IN DEVELOPING COUNTRIES HAS INCREASED STEADILY BY AN AVERAGE OF THREE MONTHS EACH YEAR.

01.12

Life Expectancy	Year	Invention / Discovery
25	1796	FIRST VACCINATION IS DEVELOPED BY EDWARD JENNER (ENGLAND); FOR THE FIRST TIME HUMANITY HAS A CHANCE TO STOP AN EPIDEMIC.
30	1818	FIRST BLOOD TRANSFUSION.
	1833	DISCOVERING THE INSIDE OF A CELL DEVELOPED BY ROBERT BROWN (ENGLAND); FOR THE FIRST TIME THERE IS AN UNDERSTANDING THAT LIFE, PHYSIOLOGY, AND GENETICS RELY ON THE UNDERSTANDING OF CELLS.
45	1840	IMMUNOLOGY IS DEVELOPED BY JAKOB HENLE (SWITZERLAND); THE IMPACT IS AN UNDERSTANDING OF BETTER WAYS OF FIGHTING DISEASE, PARTICULARLY IN RELATION TO SKIN AND ORGAN TRANSPLANTS.
	1846–7	ANESTHETICS IS DEVELOPED SEPARATELY BY IGNAZ SEMMELWEIS (HUNGARY) AND WILLIAM T. G. MORTON (USA); SURGERY CAN NOW BE CARRIED OUT WITHOUT TRAUMA ON UNCONSCIOUS PATIENTS.
	1853	EPIDEMIOLOGY IS DEVELOPED BY JOHN SNOW (ENGLAND) ALLOWING FOR CAREFUL OBSERVATION AND RECORDING THAT REVEALS WHERE SICKNESS COMES FROM; ONCE THE SOURCE OF A DISEASE CAN BE IDENTIFIED, IT CAN BE ATTACKED.
	1865	GENETICS IS DISCOVERED BY GREGOR MENDEL (CZECH REPUBLIC); MENDEL'S LAWS OF INHERITANCE ALLOWS OTHERS TO DEVELOP MODERN GENOMICS.
	1865	ANTISEPSIS IS DISCOVERED BY LORD JOSEPH LISTER (SCOTLAND); SURGERY CAN NOW BE COMPLETED WITHOUT PATIENTS BECOMING INFECTED, WHICH LEADS TO MUCH HIGHER SURVIVAL RATES.
	1877	GERMS ARE DISCOVERED BY ROBERT KOCH (POLAND), WHO ISOLATES THE *BACILLUS ANTHRACIS*; SCIENTISTS CAN NOW CULTURE, IDENTIFY, AND BEGIN TO DEFEAT DISEASE-CAUSING BACTERIA.
	1884	THE RELATIONSHIP BETWEEN A POOR DIET AND DISEASE IS MADE BY KANEHIRO TAKAI (JAPAN).
	1895	X-RAYS ARE DEVELOPED BY WILHELM CONRAD RONTGEN (GERMANY), WHO DISCOVERS THAT THE INVISIBLE RADIATION IS ABLE TO PASS THROUGH SOLID OBJECTS; THE DISCOVERY OPENS UP A NEW BRANCH OF MEDICINE, INSPIRES THE DISCOVERY OF RADIOACTIVITY, AND ALLOWS X-RAY DIFFRACTION TO BE INVENTED.
	1898	THE FIRST DRUG, ASPIRIN (ACETYLSALICYLIC ACID) IS DEVELOPED BY FELIX HOFFMANN (GERMANY).
57	1899	BAYER MARKETS ASPIRIN.
	1902	FIRST MODERN ELECTRICAL AIR-CONDITIONING UNIT IS INVENTED BY WILLIS HAVILAND CARRIER (USA).
	1921	INSULIN IS DEVELOPED BY FREDERICK BANTING AND CHARLES BEST (CANADA).
	1928	PENICILLIN (ANTIBIOTICS) IS DEVELOPED BY ALEXANDER FLEMING (ENGLAND); THE FUNGUS CALLED *PENICILLIUM NOTATUM* MAKES A CHEMICAL THAT CAN STOP THE GROWTH OF BACTERIA.
65	1930s	THE JET ENGINE IS DESIGNED INDEPENDENTLY BY FRANK WHITTLE (ENGLAND), WHO WAS THE FIRST TO REGISTER A PATENT IN 1930, AND HANS VON OHAIN (GERMANY), WHOSE JET WAS THE FIRST TO FLY IN 1939.
	1932	ALTHOUGH THE FIRST CARDIAC PACEMAKER WAS INVENTED BY MARK LIDWELL (AUSTRALIA) IN 1926, IT IS ALBERT HYMAN'S (USA) INVENTION IN 1932 THAT BECOMES UNIVERSALLY KNOWN AS THE "ARTIFICIAL PACEMAKER."
	1938	THE COMPUTER TAKES SHAPE THROUGH KONRAD ZUSE'S (GERMANY) Z1 CALCULATING MACHINE, AND WILLIAM HEWLETT AND DAVID PACKARD'S (USA) 200A AUDIO OSCILLATOR.

Life Expectancy	Year	Invention / Discovery
	1940s	RESEARCH BECOMES MORE SIGNIFICANT.
	1952	VACCINES BEGIN FOR POLIO, TUBERCULOSIS, MUMPS, AND RUBELLA. THE FIRST MECHANICAL HEART, DEVELOPED BY FOREST DEWEY DODRILL (USA), IS USED SUCCESSFULLY DURING OPEN-HEART SURGERY.
	1954	FIRST SUCCESSFUL KIDNEY TRANSPLANT.
71	1960	THE BIRTH CONTROL PILL BECOMES AVAILABLE. THE FIRST SUPERCOMPUTERS ARE INTRODUCED, PRIMARILY DESIGNED BY SEYMOUR CRAY (USA).
	1967	FIRST HEART TRANSPLANT.
	1970	MICROPROCESSOR IS DEVELOPED FOR COMMERCIAL APPLICATIONS.
	1972	FIRST COMMERCIALLY VIABLE CAT SCANNER (ENGLAND) WAS ANNOUNCED TO THE PUBLIC. ANTIDEPRESSANT DRUG PROZAC DISCOVERED, BECOMING AVAILABLE IN THE US MARKET IN 1988.
	1977	MRI DEVELOPED TO CREATE DETAILED IMAGES OF THE HUMAN BODY.
	1979	FIRST TRANSFUSION OF ARTIFICIAL BLOOD.
	1983	STOMACH ULCERS BECOME EASIER TO CURE WITH THE DISCOVERY BY BARRY MARSHALL AND ROBIN WARREN (AUSTRALIA) THAT THEY ARE CAUSED BY BACTERIA, NOT STRESS.
	1986	HIGH-TEMPERATURE SUPERCONDUCTORS ARE DISCOVERED BY ALEX MÜLLER AND JOHANNES GEORG BEDNORZ (SWITZERLAND).
77	1990	THE FIRST CONTRACEPTIVE HORMONAL IMPLANT BECOME AVAILABLE IN THE USA.
78	2009	LIFE EXPECTANCY BASED ON LATEST DATA FROM THE CDC.

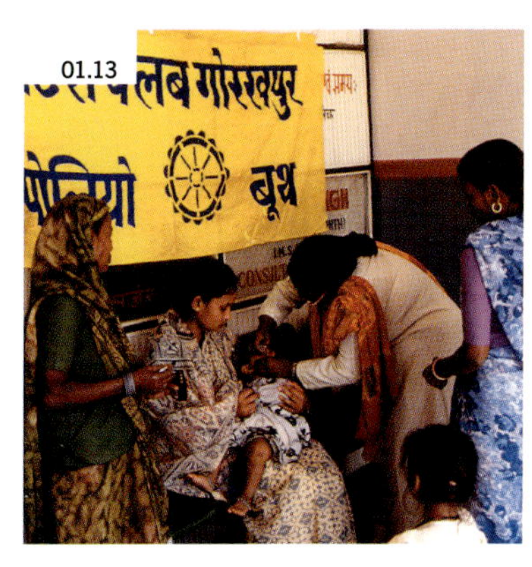

01.13 / SINCE 1952, VACCINATIONS FOR POLIO, LIKE THIS ONE BEING GIVEN TO A CHILD IN INDIA, HAVE BEEN ADMINISTERED AROUND THE WORLD. THROUGH ITS POLIOPLUS PROGRAM, ROTARY INTERNATIONAL IS A KEY ORGANIZATION IN THE GLOBAL EFFORT TO ERADICATE THE DISEASE.

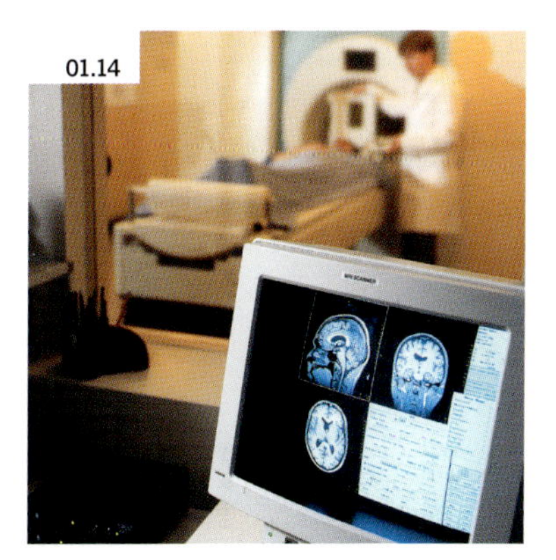

01.14 / MEDICAL IMAGING HAS COME A LONG WAY SINCE THE FIRST CAT SCAN APPEARED IN 1972.

01.12 / INVENTIONS, SCIENTIFIC DISCOVERIES, AND MEDICAL BREAKTHROUGHS HAVE HAD A HUGE IMPACT ON LIFE EXPECTANCY SINCE THE END OF THE 18TH CENTURY.

Some of the data listed in the table came from Peter Macinnis's book *100 Discoveries: The Greatest Breakthroughs in History.*[6] Macinnis correctly states: "Out there, somewhere, there is a discovery, an observation, a measurement that does not quite fit the present model for something. There is an idea, a notion, a hunch that will some day become a great discovery of science." Science builds on advances and needs. Our societies, cultures, and needs, as well as economic opportunities have clearly shaped the way science has evolved and changed.

01.15 / MATERNAL MORTALITY (1915–2000)

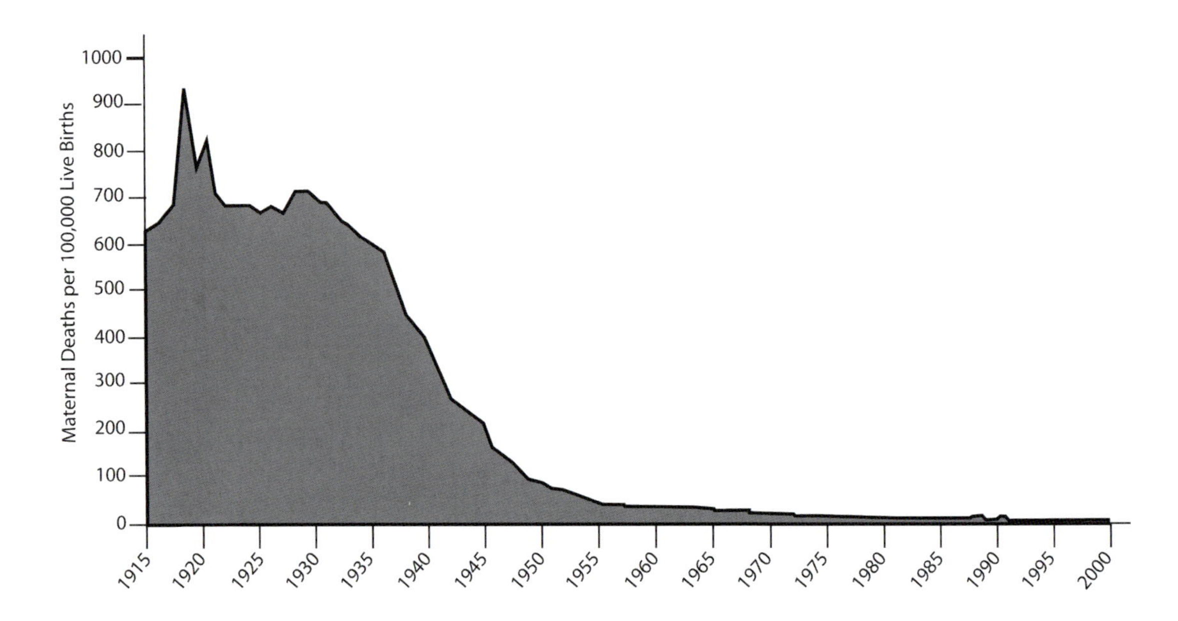

Research has also had a dramatic impact on reducing infant and maternal mortality. Much of this has a direct correlation to the better education and healthcare practices generated by effective research over the past 150 years.

Opportunities to Solve Health Problems /

Chronic Diseases /

Throughout history infectious diseases have been mankind's greatest enemy, but due to progress in infectious disease control this is no longer the case. Rather, chronic diseases have become the greatest killers.

"The great epidemics of tomorrow are unlikely to resemble those that have previously swept the world," reported the World Health Organization (WHO) in its 2005 report *Preventing Chronic Diseases: A Vital Investment.* "While the risk of outbreaks, such as a new influenza pandemic, requires constant vigilance, it is the 'invisible' epidemics of heart disease, stroke, diabetes, cancer and other chronic diseases that now require a greater commitment to research, education regarding lifestyles, and policies."[7]

Common chronic diseases include cardiovascular diseases, cancer, chronic respiratory disorders, and diabetes. Others include nueropsychiatric and sense-organ disorders, musculoskeletal and oral disorders, digestive diseases, gentourinary diseases, congenital abnormalities, and skin diseases.

Over one billion people worldwide are overweight, which has led to a sharp rise in diabetes. The problem is significant, and growing worse, in the US. As awareness of this serious problem grows, attention is becoming focused not on its cause, but on medical treatments to mitigate its consequences. The real cause of obesity is embarrassingly simple: Americans consume more calories than they need in order to maintain a healthy body weight. According to the US Department of Agriculture, the average American consumed 500 calories more per day in 2000 than in 1970. Much of this increase is explained by the doubling in the amount of food eaten outside the home from the mid-1970s to the mid-1990s, by which time restaurant and takeout food accounted for one-third of the total energy consumption. **Recent data has shown a direct correlation between diabetes and obesity to many types of cancer.**

Clearly, the outlook for Americans' health is not good when one of the key risk factors for most chronic diseases is increasing at an epidemic pace and little is being done to get at the heart of the problem. To reduce or ideally prevent some of these chronic diseases there will need

01.16 / INFANT MORTALITY (1915–2000)

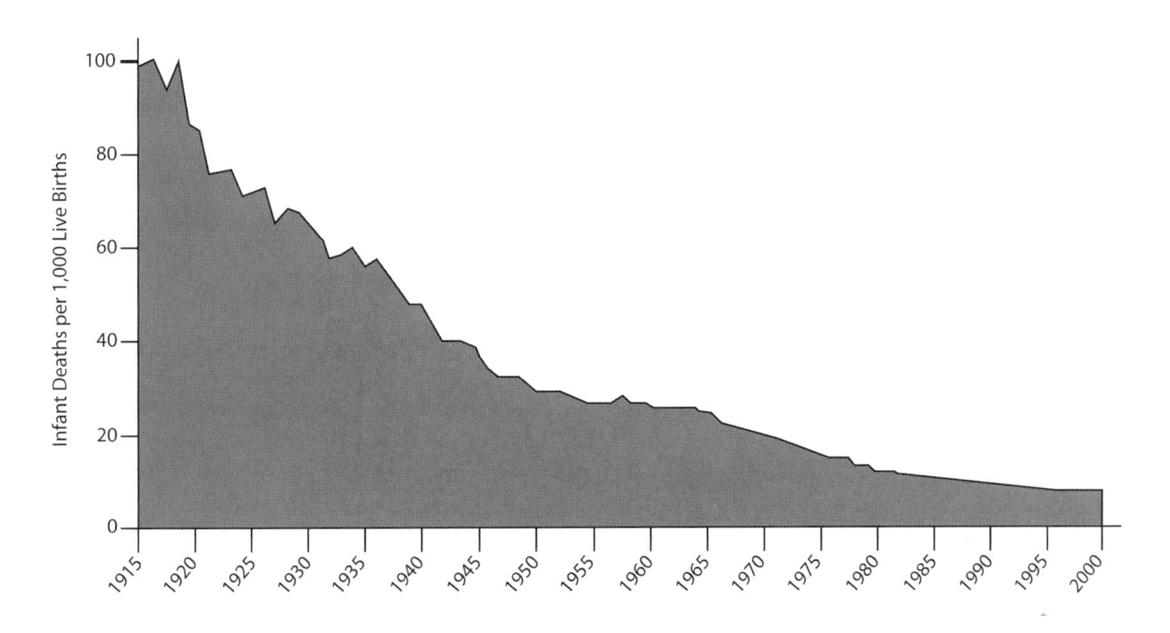

to be changes through diet and lifestyle interventions. Walking at least one mile per day for people with diabetes has cut their death rate in half. Diabetes is the fastest growing disease in the US, and used 34 percent of the Medicare budget in 2005.

Over the course of an average year 62 percent of Americans take no drugs at all. However, **three-fourths of the elderly do take drugs, half of whom take two or more that requires them to follow a daily regimen, usually for chronic conditions such as high-blood pressure, diabetes, or arthritis**. "There is simply no excuse for allowing chronic diseases to continue taking millions of lives each year when the scientific understanding of how to prevent these deaths is available now," stated the WHO in *Preventing Chronic Diseases: A Vital Investment.* [8]

And according to the WHO, research is another critical component: "Many chronic diseases would benefit from the development of new medications or medical devices. The private sector has a significant role to play in closing these gaps, as do public–private partnerships, which can invest strategically to accelerate progress with regard to specific diseases. Alternatives to insulin-delivery technologies, such as nasal sprays, could reduce the need for trained personnel, injection needles, and refrigeration, and could revolutionize the management of diabetes."

Dr. David A. Kessler, head of the US Food and Drug Administration and perhaps best known for his efforts to investigate and regulate the tobacco industry, is now focusing on the quality of food. Dr. Kessler believes food makers have a good understanding of the human brain, taste preferences, and desire to eat certain foods. In his book *The End of Overeating: Taking Control of the Insatiable American Appetite* he provides practical advice for using the science of overeating to our advantage, so that people can begin to think about food and take control of their eating habits. One of his key points is that overeating is usually not due to an absence of willpower, but a biological challenge made more difficult by the over-stimulating food environment that surrounds us. [9]

Research and collaboration are key drivers in advancing global well-being. If research discoveries continue to improve the quality of our lives over the next 150 years as they have done over the past 150 years, future generations will have much to look forward to. We now have the technology and knowledge to improve the quality and duration of lives among people of all nations. To realize the potential, however, that amazing achievement must be crowned with political commitment, social reforms, and financial support. ■

02. INTERNATIONAL COLLABORATION

International Collaboration /

As the world population continues to grow and people live longer, there will always be pressure to reduce healthcare costs. Economic pressure, technology, and globalization have given rise to international collaboration. As a result, science is being revolutionized. Freely shared knowledge is quickening the pace of advances, and problems are being addressed with a worldview of cooperation.

Collaboration is a defining trend of the early 21st century, and nowhere is it more evident than in science. **"International research collaboration is a rapidly growing component of core research activity for all countries," began a 2007 report commissioned by the UK Office of Science and Innovation.** "It enables researchers to participate in networks of cutting-edge and innovative activity [and] provides opportunities to move further and **faster by working with other leading people in their field. It is therefore unsurprising that collaborative research is also identified as contributing to some of the highest impact activity."**[1]

The US National Science Foundation holds that **"collaborative activities and international partnerships are an increasingly important means of keeping abreast of important new insights and discoveries critical to maintaining US leadership in key fields."**[2] Increased research collaboration in the industrial sector is evidenced by the growth of both research and development activities located overseas, and of cooperative arrangements between US and foreign firms.

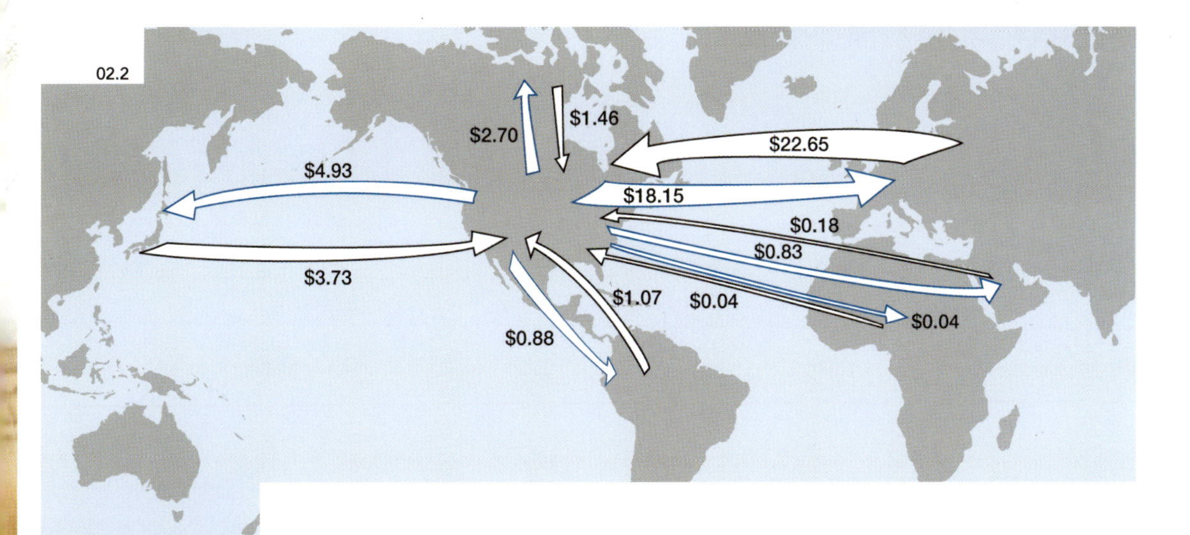

More than ever, innovation depends upon international alliances. Partners are needed to complement internal capabilities, share costs, spread market risk, expedite projects, and increase the understanding of the local market and population to produce products that will be effective and affordable.

Mobility of Research and Researchers /

The business of science is expanding its physical reach as well. Research and development functions are increasingly going abroad—and it goes both ways. While foreign-based companies seeking to conduct research still flock to the US because of available capital, brainpower, facilities, equipment, and the entrepreneurial culture, US concerns are expanding functions elsewhere. This is especially true in India and China, where some research is more cost effective, qualified professionals are plentiful, and market share has ample room to grow.

Historically, corporate America's research and development dollars were spent at labs within the US, but now the outsourcing and offshoring of research and engineering projects are skyrocketing. China and India, in particular, with low costs and large talent pools (including throngs of new graduates with engineering and science degrees), are increasingly attracting research dollars invested by US companies.

Multinationals, seeking to expand, evaluate countries on the following criteria:[3]

- Cost of labor
- Effectiveness of the national labor system
- Availability and cost of capital
- Local tax requirements
- Availability and quality of research talent
- Quality of academic research institutions
- Accessibility to transportation
- Ease in communicating in local language
- Integrity of legal system
- Potential growth of local market
- Attractiveness for employees and their families to live

The demographics of the international science and engineering labor force are gradually but profoundly changing, altering the global science and technology landscape. **Many nations have made high-skilled migration an important part of national economic strategies.**[4] Educational and work opportunities are broadening globally and the labor force is increasingly mobile. Scientists migrate in search of interesting and lucrative work, and nations are sweetening the bait to retain and attract talent.

02.1 / HEALTH WORKERS TAKING PART IN A PROGRAM TESTING FOR TRACHOMA IN TANZANIA.

02.2 / THE MAP ILLUSTRATES HOW MUCH MONEY THE USA IS INVESTING IN OTHER COUNTRIES FOR DEVELOPMENT AND RESEARCH. THE MAP ALSO DOCUMENTS HOW MUCH MONEY SOME COUNTRIES ARE INVESTING IN THE USA FOR RESEARCH.

"In the past, these countries have been a main source of internationally mobile scientific and technical talent, but recently some of them have developed programs designed to retain their highly trained personnel and to even attract people from abroad," the National Science Foundation reported in 2004. **"Because their more developed counterparts also face this issue, these trends have set up the potential for growing competition in the recruitment of foreign talent and for continuing international mobility of firms to low-cost countries with well-trained workforces."**[5]

Worldwide, the number of people with a post-secondary education—a measure of a highly educated science and engineering workforce—has skyrocketed over two decades, from about 73 million in 1980 to 194 million in 2000. Over the period, the US share of the total, which was the largest share, fell slightly, while for China and India it doubled.

The reallocation of the international scientific workforce is eroding the US's long-standing advantage. For more than half a century, the US has harbored, to its great advantage, a disproportionate percentage of the world's most highly trained science and engineering professionals, many of them foreign-born. Over the next 20 years I believe there will be more young researchers working in developing countries, many with ties to multi-national companies and global partnerships. Asian countries are becoming more successful in persuading their own skilled professionals to remain or return home and in attracting other professionals from abroad. India has lured many of its best engineers home by offering wages on par with those received by engineers in the US, *The Wall Street Journal* reported in late 2007.

The National Science Foundation reported in 2006 that: **"The number of individuals with science and engineering degrees reaching traditional retirement ages is expected to triple. If this slowdown occurs, the rapid growth in research-and-development employment and spending that the United States has experienced since World War II may not be sustainable."**[6]

Research Profiles by Country /

To maximize collaborative research opportunities and to benefit from the knowledge of others, one must understand the focus of research activities across the globe. This section looks at research activities by nation. Continents are presented alphabetically; countries are presented according to research productivity. An overview of research activities in the US is presented in Chapter 3. The global economic downturn in 2008–09 has slowed investment in research. Currently most countries are trying to maintain as many of their existing programs as possible.

Asia /

55 countries

4 billion people

17,139,000 square miles

Several countries in Asia are becoming players in the global research economy, in part due to Asia's sheer size (it has two-thirds of the world's population). Although China and India garner most of the press, research and development is burgeoning throughout the continent.

China /

Within a single generation, China has become a leading world economy. From 2000 to 2005, China's research and development expenditures increased an average of 24 percent annually. In April 2005, Chinese Premier Wen Jiabao summarized the Chinese economic strategy in stark terms: **"Science and technology are the decisive factors in the competition of comprehensive national strength."**[7]

In May of 2009 the Chinese government announced a program to attract scientific talent back to China. At a time when China has increased efforts to build a high-level science and technology industry, the increased competition for academic talent among universities is a focus in their new "One Thousand Talents Plan." This is a

02.3

02.3 / APPROXIMATE EXPENDITURE OF RESEARCH DOLLARS (BILLIONS) PER COUNTRY (2008)

national human resource program to attract back to China high-level academics who have gone overseas to work or study. These international researchers are becoming a new force in Chinese universities and are needed resources in order to grow China's science and technology skills. As China's economy is relatively strong in the current global downturn, many Chinese who are abroad view this as a good opportunity to go back and work in China.

China's progress is dramatic, but some serious hurdles remain. China remains an exceedingly poor country; its 2005 per capita annual income was only US$1,284. Beyond its weak intellectual property protections, "China faces numerous obstacles to joining the ranks of the world's innovation leaders," wrote *The Wall Street Journal* in 2006. Furthermore, research facilities are generally primitive; leading-edge equipment is found in only a few elite Chinese labs. There is also a shortage of senior-level researchers who can lead major projects. China's State Council has said it will boost research and development investment to 2 percent of gross domestic product in 2010 and to 2.5 percent by 2020. Tax breaks and other means have been outlined to meet that target. If the Chinese accomplish this, their investments will be significant and they will clearly be a world leader.

In 1998, China began a consorted effort to improve its educational system. It increased funding, reorganized institutions and made the curriculum more flexible. As a result, "natural sciences and engineering enrollment in Chinese universities grew from roughly 1.8 million students in 1995 to 5.8 million in 2003," with more than half of all undergraduate students enrolled in these fields, reports the National Science Foundation.[8] Furthermore, between 1996 and 2001, China nearly doubled its output of science and engineering PhDs reported *Electronic Business* in July 2005. By 2010, China is expected to surpass the US in numbers of science and engineering doctorates.[9]

The pharmaceutical industry, in particular, is attracted to China's well-trained scientific labor pools and its lower living and business costs. Patient enrollment in clinical trials is expected to proceed at a faster pace in India and China than in Western countries. "Costs for conducting clinical trials in China are about one-fourth of those in the US. China is beefing up standards in a bid to attract more Western companies to test drugs there," reported *Business Week* in August 2005.[10] The recent establishment of a Food and Drug Administration-like regulatory body in China is an important change, allowing Hong Kong to conduct clinical trials in China, the largest market in the world.

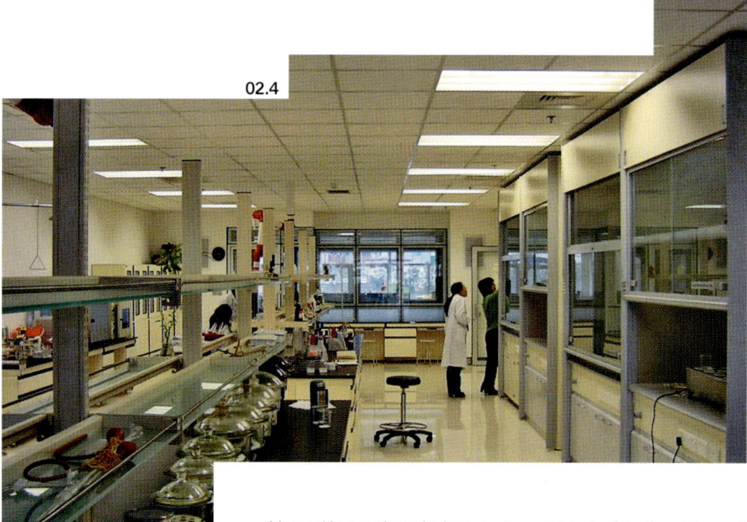
02.4

Hong Kong already has a strong pool of trained investigators and history of earning ultra-fast regulatory approvals. Collaboration between Hong Kong and China will be mutually beneficial. However, China will become a leading greenhouse gas producer by 2010. According to *Wired* magazine, "environmental degradation has become a drag on China's development. The government revealed last year that environmental damage costs the economy $200 billion a year, a full 10 percent of China's GDP."[11] The country realizes that no economic policy can be successful without a sound environmental policy. The cost to public heath and quality of life may be even greater, *Wired* stated. "Over cultivation, overgrazing, and massive timber consumption have turned a quarter of China's land into desert. Over 400 million Chinese drink contaminated water ... The government figures that 300,000 people die prematurely each year from polluted air."

On a trip to China in June 2009 I read articles each day in the *China Daily* that focused on economic and environmental issues. In the June 2 edition, Liu Qi, deputy director of the national energy administration, stated: "We are aiming to make the new energy industry a new engine of development." The article went on to explain that the first phase of the **Chinese Energy Program covers a shift in three years to nuclear, solar, wind, biomass power, and clean coal technologies—with investment opportunities equal to approximately US$440 billion**. Phase two will be completed by 2020 and entail much more investment. Currently wind power generates only 2 percent of the country's energy but by 2020 wind power will likely surpass nuclear power as China's third source of electricity, after solar power and hydropower. The new energy program will be a driving force in growth in China once the 2009 stimulus plan stabilizes the overall economy.

"The Chinese advantage is that when they decide something, they can do very dramatic things," energy analyst Jim Brock told *Wired* magazine. "In 2000, they took 26,000 heavily polluting minibuses off the road in a week [in Beijing]. They cut the pollution by 6 percent just by saying we don't want these cars on the road. Try that in the United States—it wouldn't work."

I believe it is highly likely that China will be one of the key leaders in addressing today's energy problems in a very proactive manner. Expect to see China construct very large and successful wind and solar farms over the next decade. I started designing research buildings in China in 1998. Since that time I have seen significant growth and great improvements in the quality of the construction, which now is close to the quality of work found in Europe and the US. The biggest problem is getting reliable mechanical, electrical, and plumbing systems, although it is only a matter of time before there are improvements here as well.

Japan /

Japan once accounted for the lion's share of high-technology output in Asia, but its long-term output has grown more slowly than that of other Asian economies and dropped sharply after 2000, reported the National Science Foundation. Japan's share of Asia's total high-technology output declined from 77 percent in 1990 to 50 percent in 2000, and then to 38 percent in 2003.[12]

Japan leads the world in robotics and shares dominance in nanotechnology, along with the US, Germany, South Korea, and China. Other technological strengths include electronics, machinery, optics, chemicals, semiconductors, and metals, and the country houses six of the world's largest automobile manufacturers. Japan has also made headway into aerospace research and space exploration. The Japan Aerospace Exploration Agency conducts space and planetary research, aviation research, and development of rockets and satellites. Japan's success in rebuilding its economy after the Second World War is a textbook example of the power of technological innovation to stimulate social development. By the 1980s, the country was an economic juggernaut, boldly threatening US dominance.

"Japan's competitiveness seemed unassailable, with a strong domination in economic dynamism, industrial efficiency and innovation," writes Switzerland's International Institute for Management Development (IMD), ranked second internationally among business schools by *The Wall Street Journal.*[13] "Then all hell broke loose: the stock market went into reverse in 1989, land prices collapsed in 1992, credit cooperatives and regional banks came under attack in 1994, large banks teetered on the edge of bankruptcy in 1997 and a major credit crunch occurred in 1998. Does this ring a bell?" asks IMD, referencing the economic turmoil in the US in 2008.

Japan has the political will to push through reforms to improve competitiveness. Continuing the momentum initiated by his predecessor, Prime Minister Shinzo Abe, who actively promotes innovation as being important to Japan, chairs the Council for Science & Technology Policy once a month.

The Japanese government has an ambitious target of growing its biotech industry to US$252 billion and creating 1,000 new biotech companies by 2010. Japanese biotech companies have expertise in bioinformatics, manufacturing, and genetics, and have formed several partnerships with multinational companies. The country is also creating more smart research partnerships between private industry and universities.

02.4 / JAHWA'S NEW PHARMACEUTICAL RESEARCH LABS WERE COMPLETED IN 2001. THIS WAS THE AUTHOR'S FIRST LABORATORY PROJECT.

Japan has historically excelled in electronics and information technology. Banking on nanotechnology as the key to its economic future, the government founded the Expert Group on Nanotechnology under the Japan Federation of Economic Organizations Committee on Industrial Technology.

Singapore /

The tiny island-state of Singapore is the 17th wealthiest country in the world in terms of GDP per capita. Singapore has committed billions of dollars to develop education, technology, and infrastructures to grow its new economy. The government is funding innovative, interdisciplinary research projects, particularly in rapidly evolving areas such as genomics, proteomics, nanotechnology, tissue engineering, bioinformatics, and biomedical engineering. Singapore has also created a world-class center for scientific research and development.

Although Singapore is notably conservative on most social issues (including a ban on most types of chewing gum), it has addressed the ethical, legal, and social issues arising from scientific research. The country believes that biomedical research, in particular, must be well-founded on ethics and practices that are acceptable to society and the international community. The government-appointed Singapore Bioethics Advisory Committee (BAC) developed policy recommendations through extensive study, consultation (with the public, professional and religious organizations, and healthcare and research institutions) and deliberation. Many BAC recommendations have been adopted.

Since launching its Biomedical Sciences Initiative in 2000, Singapore's biomedical industrial output almost quadrupled, from US$4.36 billion to US$16.6 billion by the end of 2006. The biomedical-sciences industry accounts for over 6 percent of Singapore's GDP, and in manufacturing alone it has provided approximately 10,000 high-value jobs. The biopharmaceuticals sector has diversified to biologics and vaccines. In medical technology, Singapore produces high-value products such as complex instrumentation systems and tissue heart valves.

Singapore has lured big drug companies using the same combination of tax incentives that drew leading electronics makers. Some 40 biomedical science companies have research operations in Singapore, including pharmaceutical companies such as GlaxoSmithKline, Eli Lilly, Novartis, and Takeda.

To entice companies to go beyond making drugs and to invest in basic drug research and development, Singapore pays up to 30 percent of the building costs for new facilities. As of 2006, at least 30 companies had responded, "including the Swiss drug giant Novartis, which has opened an institute here to develop drugs to fight tuberculosis and the dengue virus," reported the *The New York Times*. In anticipation of accelerated growth, particularly in the biomedical sciences, the Singapore government set aside nearly 500 acres in 2001 to create a science Mecca called Biopolis in the suburb of Buona Vista. Completed in 2004, Biopolis houses educational institutes, info-communication technology organizations, media and bioscience firms (including a new US$20 million multinational biomedical laboratory), and venture capital firms. An estimated 150,000 people will eventually be employed there.

Singapore is also building a new research town, the Campus for Research Excellence and Technological Enterprise (CREATE) (discussed in more detail in Chapter 10). CREATE's first of many world-class research centers will be a joint venture of the Massachusetts Institute of Technology (MIT) and the National Research Foundation of Singapore. The Singapore-MIT Alliance for Research and Technology (SMART) Center, to open by 2010,

02.5

will be MIT's first such research center outside of Cambridge, Massachusetts, and its largest international research endeavor.

Singapore is concurrently investing in the development of future biomedical research talent. Postgraduate education scholarships have been funded and The Singapore Agency for Science, Technology and Research (A*STAR) provides scholarships to the region's best students to pursue PhDs in the biomedical sciences, physical sciences, and engineering. Beyond the PhD students in regular programs in Singapore universities, more than 600 A*STAR scholars pursued studies in Singapore, the US, and Europe in 2007.

It costs up to US$600,000 to finance the education of each student from a basic university degree to a PhD, eight years of study that the scholar must repay with six years of service to the government. **Singapore is clearly a model of growth and development for other countries to emulate in the 21st century**.

India /

As part of the BioAsia program, several Indian states are establishing biotech parks near major universities. "Over the next five years, India's burgeoning biotech industry will generate (US)$5 billion in revenues and more than one million jobs, according to a report recently released by Ernst and Young."[14] The IT revolution is creating jobs and changing lives, especially for millions of educated young women.

02.5 / AXEL ULRICH OF GERMANY, YOSHIAKI ITO OF JAPAN, AND DAVID LANE OF THE UK, ALL TOP CANCER EXPERTS FROM COUNTRIES WHO HAVE BEEN DRAWN TO BIOPOLIS.

The state of Maharashtra is a hotbed of science and technology. Its two technology parks—one at Hinzewadi near Pune, and a second for agriculture biotech in Aurangabad—are a few hours drive from Mumbai. More than 40 percent of India's pharmaceutical sales and 30 percent of all patents filed in India are from companies and academic institutions in

02.6

Maharashtra. New product development in India focuses on the research and development of pharmaceuticals to combat TB, HIV/AIDS, malaria, cholera, cancer, typhoid, and heart disease. Much has been said about the impact of China's high rate of economic growth on the global economy, but India is also a major player and is capable of tackling many of the world's common diseases.

The US venture capital community is taking a cautious approach toward investing in biotech in India as it believes the country is not yet equipped to follow the US biotechnology model. Among other reasons, India suffers from deficiencies in its biological testing capabilities and lacks sufficient research-and-development management talent. One business model that could overcome these hurdles is a company that creates the research-and-development management, business development, and financing in the US, and relegates its office work to India. This could open the doors for US technology and financing while utilizing India's potential to develop, trial, and manufacture drugs.

Alan Levi, CEO of Pfizer (New York City) global research and development, emphasized that globalization of research and development is essential for controlling costs and accelerating development. India offers not only scientific talent but also many customers for new drugs.

Other roadblocks to doing business in India include weak intellectual property protection and limited laboratories doing basic research (due to lack of grant programs such as NIH Small Business Innovation Research funds). A clear vision and a strong commitment are needed from the Indian government to resolve these issues.

South Korea /

South Korea is now ranked seventh in the world in research-and-development spending among its companies. "The Korean Government has taken additional bold actions to strengthen the nation's science and technology capabilities. It plans to build a 'Science and Technology-oriented Society.' As a start, the science and technology Minister is now a Vice Prime Minister position. The National Innovation System will be reworked to improve university research, seek globalization and to concentrate national science and technology resources on the development of 10 new growth industries. To facilitate commercialization of public research and development results, the technology transfer budget was more than tripled."[15]

02.6 / RESEARCH BUILDING ON THE CAMPUS OF KING SAUD BIN ABDULAZIZ UNIVERSITY FOR HEALTH SCIENCES IN RIYADH, SAUDI ARABIA.

Saudi Arabia /

Saudi Arabia, and many other Middle-Eastern countries, are building major infrastructure projects with the money they have made from oil. I have been involved in the design of three new medical schools—in Al Hasa, Jeddah, and Riyadh.

Europe /

55 countries

830 million people

3,930,000 square miles

Europe is a likely beneficiary of the new science world order. Already equipped with a world-class science base, Europe is strengthening its capabilities in stem cell research, genomics, proteomics, neuroscience, and nanotechnology. The European Union and its member states are increasing research-and-development investment to 3 percent of GDP by 2010.

The majority of European countries have begun to introduce national measures to stimulate private-sector spending in research and development. "Of particular importance in this respect are tax incentives, which eight member states have already put into place and account for 13 percent of direct research investment in these countries," reported EurActiv.com.[16]

"Europe's citizens are concerned by important issues ranging from climate change and the depletion of non-renewable resources to demographic change and emerging security needs, which call for collective action to safeguard the European way of life that combines economic prosperity with solidarity. These legitimate concerns must be turned into an opportunity to enhance Europe's global economic competitiveness. The quicker it can react, the higher the chance of success and the greater prospect that its approach will serve as a global model. From the protection of the environment through eco-innovation to the improvement of individual well-being through more intelligent infrastructure provision, the Commission is convinced that innovation in a broad sense is one of the main answers to citizens' material concerns about their future," stated the Commission of the European Communities in the report "Putting Knowledge Into Practice: A Broad-Based Innovation Strategy for the EU."[17]

United Kingdom (UK) /

The UK, small as it is, has ranked as a leader in research for centuries. Building on the priorities set out in the country's Science and Innovation Investment Framework 2004–2014, the government seeks to optimize science and innovation by creating the best possible environment. A world-class science base to connect with business, financial incentives, and other support mechanisms will help create new knowledge-based firms that take advantage of commercial opportunities arising from research.

France /

To gain critical mass, France directs its research and development money toward clusters. **"We can't have every corner of Europe, let alone every corner of France, being a leading center for research and industry,"** says Dr. Phillippe Pouletty, managing director and co-founder of Life Sciences. **"Political leaders have to choose not to spread resources too thin."**[18] France's spending is growing sluggishly, hovering around 2.5 percent of GDP, half of it from industry. By 2012, France hopes to boost financial incentives to state-run laboratories with private-sector funding. The initiative will brand participating research centers as "Carnot" laboratories (after a French statesman and scientist), and is modeled on the German network of public-private funded Fraunhofer research institutes.

Germany /

German research has traditionally been strong in the areas of mechanical engineering, chemistry, medicine, physics, and mathematics. Today, nanotechnology and biotechnology are creating much excitement in this self-coined "land of ideas," which trails only the USA and the UK in the number of Nobel Prizes earned. With more than 250,000 scientists and investigators, Germany is the world's third-largest "country of researchers." Germany has positioned itself as a world player in three interrelated technologies that will shape the 21st century: nanotechnology, biotechnology, and information technology. Research and innovation in the field of nanobiotechnology requires collaboration between life scientists, physical scientists, and engineers.

Ireland /

Ireland has a strong enterprise base and the potential to sharply increase its research and development capability. It also has a growing public research base. In the 1990s Ireland pushed research and the GDP per capita was 69 percent of the European Union average, but by 2003 it had reached 136 percent. Ireland's unemployment fell from 17 percent to 4 percent over the same period. The main reason for this success was that Ireland courted multinational companies by offering an excellent tax rate. Most of the world's top pharmaceutical companies now have manufacturing facilities in Ireland.

Ireland's future economic well-being will be determined by how well it promotes innovation and a culture of entrepreneurship among researchers, and fosters research-and-development partnerships between private industry and academia.

Lucent Technologies' Bell Labs, one of the world's most eminent research institutions, offers a great example of a partnership between private industry and a multinational company. Lucent's Center for Telecommunications Value-Chain-Driven Research, established in partnership with Trinity College Dublin, will undertake research aimed at realizing the next generation of telecommunications networks. Georgia Tech Research Institute's new Irish operation will be a critical component of Ireland's research-and-development infrastructure, and will help to reinforce collaboration with academia around the world.

Finland /

In 2005, the Finns put 3.5 percent of their domestic product into research and development, second in the world to Sweden. **Finland is often used as a study of smart long-term strategic approaches to research. In the 1970s the country's leaders agreed to focus on research-and-development funding for electronics, biotechnology, and materials technology. From 1980 to 2003 electronic exports grew from 4 percent to 33 percent of all exports.**[19] Tekes, the national technology agency, supports both basic and applied research, granting about 40 percent of its funds to universities and other research institutions, and 60 percent to businesses.

Africa /

53 countries

1 billion people

11,688,545 square miles

Due to extreme poverty, Africa has neither the technology nor the skilled workforce to effectively participate in, and reap the benefits of, globalization. As globalization and technology reduce the economic gap between India and China and the advanced industrial countries, the gap between Africa and the rest of the world is increasing. Africa cannot be left behind. To raise the level of capital per person and eventually break the poverty cycle in Africa, donor-backed investments are needed. "When the capital stock per person is high enough, the economy becomes productive enough to meet basic needs," states economist and Columbia University professor Jeffrey Sachs.[20]

First there should be investment in primary education and healthcare. Once those have been established an infrastructure for science—well-financed universities, laboratories, a critical mass of research funding, and collegial support—will have to be built. The commitment must be long term; it may take up to 20 years to realize significant results. There are also other areas outside

02.7

of the research industry that will need to be developed but history has shown that many countries have seen economic growth based largely on research development. "The general lesson of successful economies is that governments are wise to stick mainly to general kinds of investments—schools, clinics, roads, basic research—and to **leave highly specialized business investments to the private sector**." explains Sachs.[21]

Not all of Africa is engulfed in dire poverty. South Africa, wealthy compared to the sub-Saharan region, has a well-developed educational system tied into a strong research base. Research takes place primarily in engineering or natural sciences, followed by medical and health sciences.[22] "In 2004, the National Research Foundation launched South Africa's first six research centers of excellence. In addition to facilitating research and training, the centers promote knowledge sharing and negotiate partnerships. These virtual or physical centers pull together existing resources to enable researchers to collaborate across disciplines and institutions on long-term projects that are locally relevant and internationally competitive," reports the International Education Association of South Africa.[23] The seventh research center of excellence was launched in 2006, and 13 more centers have been planned. The seven institutions focus on the following areas of research:

- Biomedical TB Research (co-hosted by the universities of Stellenbosch and the Witwatersrand)
- Invasion Biology (hosted by Stellenbosch University)
- Strong Materials (hosted by the University of the Witwatersrand)
- Birds as Keys to Biodiversity Conservation at the Percy FitzPatrick Institute (hosted by the University of Cape Town)
- Catalysis (hosted by the University of Cape Town)
- Tree Health Biotechnology at FABI (hosted by the University of Pretoria)
- The Centre of Excellence in Epidemiological Modelling and Analysis (hosted by Stellenbosch University).

Libya is planning to build 18 universities over the next few years. This is an example of an oil-rich country investing in its infrastructure in order to take better care of its people.

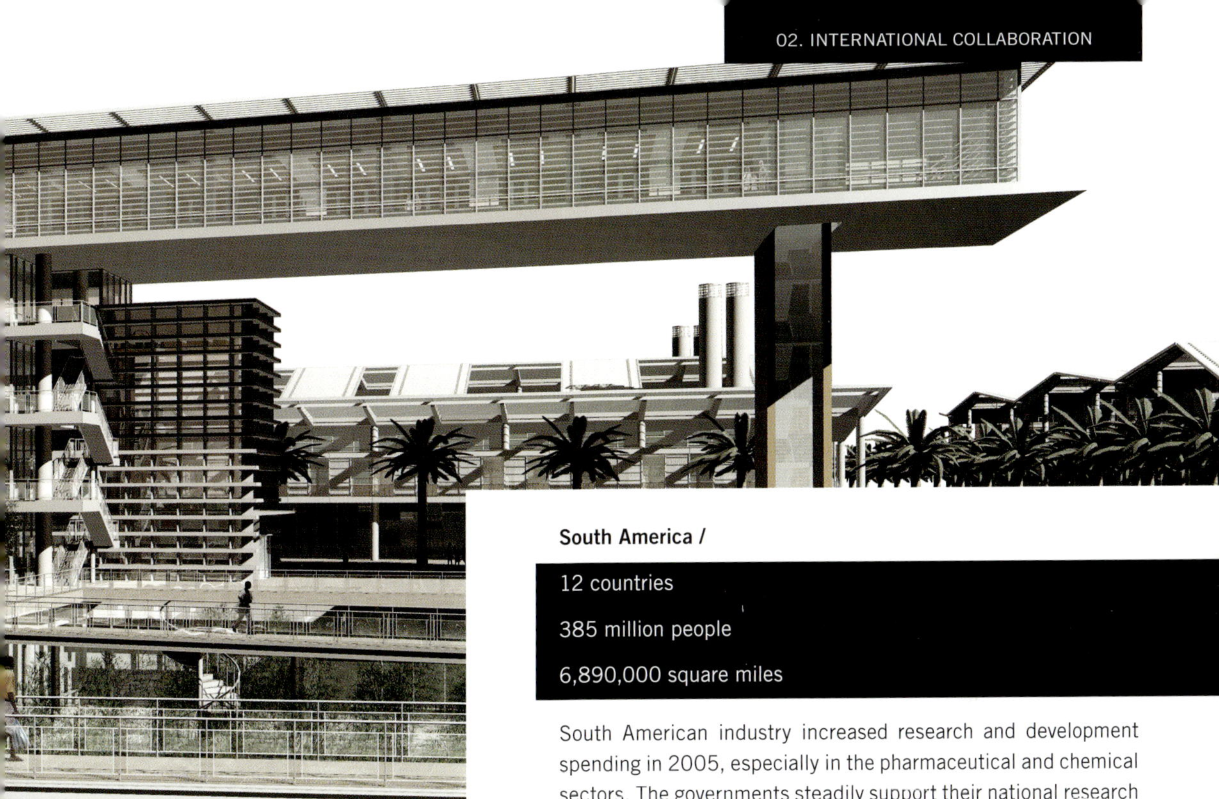

South America /

12 countries

385 million people

6,890,000 square miles

South American industry increased research and development spending in 2005, especially in the pharmaceutical and chemical sectors. The governments steadily support their national research laboratories, but at a relatively low level, and the government does not show much interest in funding technology research. However, interaction between the research laboratories and industry is increasing. Some regional innovation strategies have encouraged technology clusters. Private industry is seeking to build an international cooperation program for research, technology, and innovation such as the Framework Program for the Americas, but funding has not been identified. Joint programs with the US and Canada are also being sought. The few examples of international cooperative projects, particularly in biotechnology, are still primarily academic.

North America /

38 countries or territories

529 million people

9,540,000 square miles

Canada /

Canada hopes to rank among the top five countries in the world in research activity by 2010 (it was ranked eighth in 2003). By 2010, Canada is committed to at least doubling its investment in research and development. In addition, the government has signed an agreement with the Association of Universities and Colleges of Canada.

The US is discussed in detail in the next chapter.

02.7 / LIBRARY BUILDING ON THE CAMPUS OF UNIVERSIDADE AGOSTINHO NETO IN LUANDA, ANGOLA.

02.8

		GDP	R&D Expenditure	R&D Expenditure	Population	R&D Expenditure
	Country	2007 Source: The Economist 2010	% of GDP 2006 Source: The Economist 2010	GDP/ R&D Expenditure	2007 Source: The Economist 2010	Per person
1	United States	$13,751,000,000,000	2.61%	$343,000,000,000	303,900,000	$1,128
2	Japan	$4,384,000,000,000	3.39%	$148,400,000,000	128,300,000	$1,156
3	Germany	$3,317,000,000,000	2.53%	$73,800,000,000	82,700,000	$892
4	China	$3,206,000,000,000	1.36%	$37,700,000,000	1,331,400,000	$28
5	United Kingdom	$2,772,000,000,000	1.76%	$42,700,000,000	60,000,000	$711
6	France	$2,590,000,000,000	2.09%	$47,500,000,000	60,900,000	$779
7	Italy	$2,102,000,000,000	1.11%	$19,400,000,000	58,200,000	$333
8	Spain	$1,436,000,000,000	1.20%	$14,800,000,000	43,600,000	$339
9	Canada	$1,330,000,000,000	1.95%	$25,000,000,000	32,900,000	$759
10	Brazil	$1,313,000,000,000	1.04%	$7,300,000,000	191,300,000	$38
11	Russia	$1,290,000,000,000	1.28%	$10,600,000,000	141,900,000	$74
12	India	$1,177,000,000,000	0.8%	$9,886,800,000	1,135,600,000	$9
13	Mexico	$1,023,000,000,000	0.40%	$4,092,000,000	109,600,000	$37
14	South Korea	$970,000,000,000	3.22%	$28,600,000,000	48,100,000	$594
15	Australia	$821,000,000,000	1.84%	$11,800,000,000	20,600,000	$572
16	Netherlands	$766,000,000,000	1.65%	$11,200,000,000	16,400,000	$682
17	Turkey	$656,000,000,000	0.66%	4,239,600,000	75,200,000	$58
18	Sweden	$464,000,000,000	3.73%	$14,700,000,000	8,900,000	$1651
19	Belgium	$453,000,000,000	1.82%	$7,300,000,000	10,300,000	$708
20	Indonesia	$433,000,000,000			228,100,000	

02.8 / CHART COMPARING R&D EXPENDITURES BASED ON GROSS DOMESTIC PRODUCT

Comparison of Research Investments by Country /

As multinational pharmaceutical companies continue to expand globally and developing countries continue to expand in the research industry, there should be some thoughtful discussion of implementing research in developing countries. For example, countries such as China, India, and Puerto Rico have constructed manufacturing facilities to produce drugs and vaccines to Good Management Practices (GMP). The problem is that employees who have been hired locally, especially at the management level, do not have the experience to run the facilities. There is then a risk that the facilities won't pass inspection and won't be certified by appropriate government agencies. To resolve this problem, at significant cost, qualified managers need to be brought in from Europe or the US. What is beginning to happen is that some facilities in developing countries are being closed down to minimize risk, and the work is going back to the US and Europe. Vivarium (animal) facilities, which are operated for multinational pharmaceutical companies, are a good example of this; if one vivarium facility does not get approved for Association for Assessment and Accreditation Laboratory Animal Care (AAALAC) then all vivariums within that company are not AAALAC accredited.

Countries all over the world are striving for economic security through innovation. Those that recognize, understand, and plan for a global economy are most likely to thrive. As Bill Gates says, "**It's about expertise. The developing world has a huge shortage of expertise.**" The chart opposite, developed from data provided in *The Economist Pocket World in Figures 2010*, shows research-and-development expenditure by country. The USA leads the world in the amount of money invested in research, followed by Japan, Germany, the UK, France, and China. In research expenditure as a percentage of a country's GDP, Israel leads with 4.35 percent, followed by Sweden with 3.73 percent, Finland with 3.44 percent, Japan with 3.39 percent, and the US with 2.61 percent.

China spends a mere US$28 per person each year on research, while India spends a meager US$9 per person each year. The growth of these countries will have an impact around the world, however, China and India will need to invest US$200 per person each year on research (compared to more than US$1,000 by the USA and a handful of other countries), or more than 10 times their current investment, in order to equal the research and development expenditure of the US today.

Note that almost all of the countries listed in the Top 20 for GDP investments have major programs emphasizing research through the implementation of the following initiatives:

- Focusing on innovation and collaboration
- Increasing government funding
- Supporting private industry through partnership and tax incentives.

The initiatives that are well thought out and strategic with continued financial support behind them should see the highest rate of return on investment. Ideally all countries will benefit from the research success of others. ■

03. RESEARCH IN THE USA

Federal Funding /

The prosperity the US enjoys today is largely the result of private and public investments made in research and development since the end of the Second World War. Most industries view research and development as something they must do to stay one step ahead of the competition. The US industry was responsible for 370 new drugs and vaccines in the 1990s, half of all pharmaceutical innovation in the world. The US government continues to significantly lead the world in research spending, but its lead is shrinking.

The federal government recognizes that investment in research and education is needed to maintain leadership in the global economy, but in the past few years, funding increases have gone almost exclusively to defense-weapons development and homeland-security research and development. In his January 2006 State of the Union address, President George W. Bush launched the American Competitiveness Initiative and proposed an increase of about $910 million in yearly funding for the National Science Foundation, the Department of Energy's Office of Science and the Department of Commerce's National Institute of Standards and Technology. Longer term, the president proposed that these agencies double their budgets over a 10-year period. Due to Department of Defense funding for the war and a growing national deficit, however, the ensuing federal budgets have fallen far short of the funding needed to realize the goals of the American Competitiveness Initiative. In the 2008 fiscal-year budget, federal investment in research and development declined in real terms for the fourth year in a row.

The Obama administration is bringing about some changes. **Allocations of the American Recovery and Reinvestment Act funding, agreed to in 2009, increased federal research-and-development investment to a record $165.4 billion.** The 2009 budgets of key programs were boosted not only for their potential contribution to economic recovery but also because science and technology can help reorient the US economy through strategic investments in clean energy, broadband, healthcare information technology, and education. These laws are critical down payments toward doubling federal investments in key science agencies over a decade, meeting a commitment by President Obama to invest $150 billion during the next 10 years in a clean energy future, and enhancing America's capacity to understand the dimensions of climate change and respond to them effectively.

The new administration has doubled federal investments in three key basic research agencies—the National Science Foundation (NSF), Department of Energy's (DOE) Office of Science, and the laboratories of the Department of Commerce's National Institute of Standards and Technology (NIST). The President's Plan for Science and Innovation and the America COMPETES Act have identified NSF, DOE, and NIST as key to US prosperity and to preserving its place as the world leader in science and technology. Investment in clean-energy research and development will drive a new energy economy that reduces dependence on oil, creates green jobs, and reduces the impact of climate change.

Research and development funding will support renewable energy and energy-efficiency technologies such as advanced batteries, solid-state lighting, solar biomass, and geothermal and wind power. The 2010 budget also supports the development and testing of carbon capture-and-storage technologies that will reduce carbon emissions, and basic research to support transformational discoveries and accelerate solutions in the development of clean energy.

In real terms, the 2009 enacted level and 2010 budget are among the two largest research-and development investments in history. The American Recovery and Reinvestment Act of 2009 provides unprecedented federal support for research and development facilities and capital equipment totaling $8.2 billion. This funding includes support for the construction and renovation of laboratory facilities at government research centers, contractor-operated national research centers, and academic institutions, as well as providing funding for the purchase of major research instrumentation.

Private Sector Investments /

In 1897 Felix Hoffmann discovered the first known drug—aspirin. A few years later, Paul Ehrlich and John Langley defined the fundamental objective for drug discovery that remains current today: to find a molecule (a drug) that binds selectively to a target (receptor, enzyme, etc.) in the body to trigger a desired biological effect.[1]

Developing successful new drugs is a lengthy, scientifically complex, and expensive process. From the start of research to obtaining a patent, drug development typically takes from 10 to 15 years. A 2006 study by the Federal Trade Commission's Bureau of Economics states that for drugs entering human clinical trials for the first time between 1989 and 2002, the estimated cost per new drug was $868 million, although estimates vary from around $500 million to more than $2 billion, it added.[2] Although cost estimates vary, all agree that resources must be better utilized to lower the cost and time involved in drug development. In short, we must make a "science" out of science by improving productivity and efficiency. These ideas are discussed in more detail in Chapter 08.

03.1 / THE NATIONAL BIODEFENSE ANALYSIS AND COUNTERMEASURES CENTER AT FORT DETRICK, FREDERICK, MARYLAND.

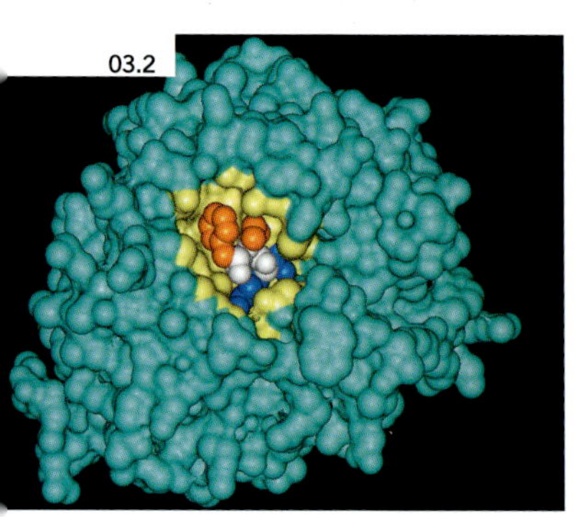

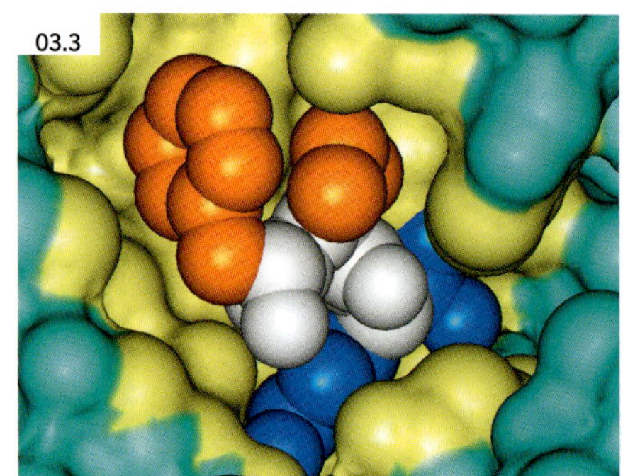

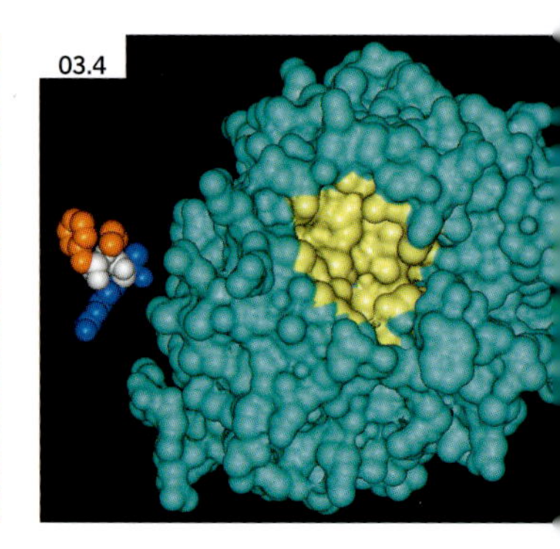

Drug Development /

There are three main stages of drug development—drug discovery, preclinical development and testing, and clinical trials. We will take a look at each one individually.

Drug Discovery /

Drug discovery involves four phases: target identification, target validation, lead identification, and lead optimization. In the first phase, target identification, scientists must identify the molecular basis of the design of the disease. Many refer to the drug discovery process as "lock and key." The drug is the key and the lock is the receptor or enzyme that is the biological reaction to the drug. If the key fits in the lock completely it will block the active site of the enzyme, thus killing the virus. If the key does not bind to the lock, or binds differently than expected, there is the potential of another discovery thus sending the research off in a different direction and starting another cycle of design. Many cycles are usually required to get a successful "hit". Much of this work is completed using 3D computer modeling.

The computer images above are a result of Tim Maher's work at the Oklahoma Medical Research Foundation (OMRF). His work tests ideas with the 3D models, gets the ideas close to being finalized, and then the ideas are developed further in the laboratory.

Pharmaceutical companies and small startups commonly collaborate with universities and public agencies to bring together a cross section of chemists, biologists, bioinformatists, and others to work closely on a specific problem. Some target identifications can be immediate but it typically takes three to four years to develop a group of potential targets to explore in more detail. Usually it takes one or two years to crystallize and determine the structure of the disease. Once the structure is determined the inhibitor (lock) can be tested and designed. Often this process involves the chance encounter of a collision of enzymes. Most enzymes are made of proteins.

The next six months are the target validation phase. The team studies and challenges each potential target to decide which ones, if any, should continue to be tested. About one year is then devoted to the lead identification process, whereby the team screens and tests thousands of possible molecules looking for a "hit"—a molecule that binds to a target and alters its action. The process continues in lead optimization phase to determine if there are variations of the molecule that might have better properties. The potential solution is then checked with patent lawyers to make sure the solution has not already been discovered and patented.

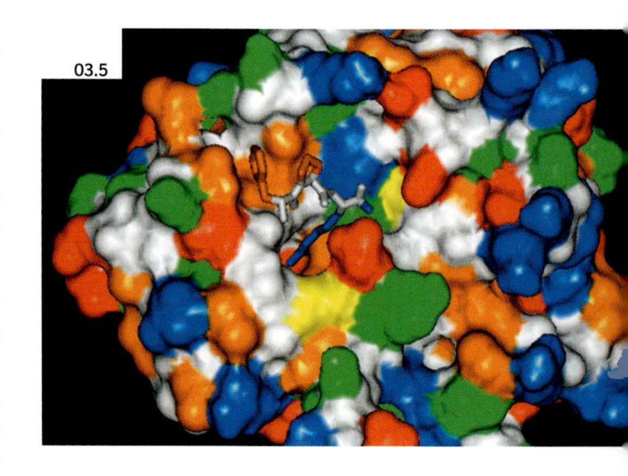

03.6

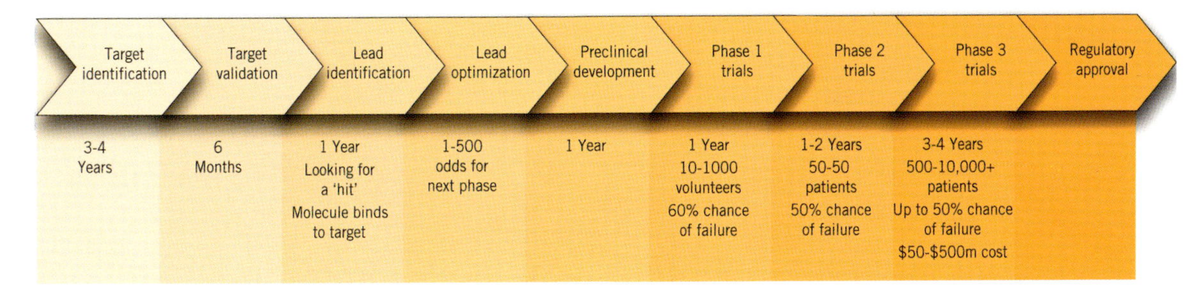

At this stage the team has a compound that they know relatively little about, except that it effectively binds to the target. At this phase of the drug-discovery process, the odds are about 1 in 5,000 that the molecule will work and be turned into a commercially viable solution. This phase takes approximately one year.

Preclinical Development and Testing /

After identifying a lead compound, researchers conduct preclinical studies—a series of in vitro and in vivo (animal) experiments—to determine if the molecule is worth pursuing in human clinical trials. This phase typically takes a year.

Clinical Trials /

If the molecule offers a potential solution, a developer will submit an Investigational New Drug (IND) application to the Food and Drug Administration (FDA). Even for the most promising drugs the road ahead is still rife with opportunities for failure as the drug undergoes toxicology and efficacy testing. The largest cost in drug development is when negative side effects are discovered.

Historically, even after an IND application is filed to begin Phase 1 of human clinical testing, there is about a 60 percent chance that the drug will fail before reaching the first phase of clinical trials. A drug that makes it to Phase 2 has only a 50 percent chance of getting to Phase 3. And even when a drug advances to Phase 3 trials, the probability of failure can be as high as 50 percent. Thus, of a portfolio of 10 new drug candidates beginning clinical trials, seven would typically make it to Phase 2, three would make it to Phase 3, and one or two products would ultimately be approved for commercial sale.

Phase 1 clinical trials try to determine dosing, document how a drug is metabolized and excreted, and identify acute side effects. Usually, a small number of healthy volunteers (between 20 and 80) are tested to assess the drug's safety. This phase takes about a year. A new business model has developed to rule out drug candidates earlier in Phase 1 to save money in the Phases 2 and 3 clinical trials. The FDA starts its review process at Phase 1 clinical trials and continues through to the final regulatory approval.

Phase 2 clinical trials involve from 100 to 300 participants who have the disease or condition that the product could potentially treat. Researchers seek to gather further safety data and preliminary evidence of the drug's efficacy. This stage lasts from one to two years.

03.2, 03.3, 03.4 / THREE EXAMPLES OF A 3-D MODEL FOCUS ON THE "LOCK AND KEY" EFFECT.

03.5 / THE MOLECULES ARE THE COLORFUL SPHERES PACKED TOGETHER, AND THE INHIBITOR IS THE STICK FIGURE. THIS 3D MODEL IS WORK GENERATED BY RESEARCHERS AT OMRF THAT WHILE WORKING ON THE DRUG XIGRIS TO TREAT SEPSIS, OR SEPTIC SHOCK. THE PROJECT WAS A PARTNERSHIP BETWEEN OMRF AND ELI LILLY.

03.6 / DRUG DEVELOPMENT FROM TARGET IDENTIFICATION TO REGULATORY APPROVAL.

In Phase 3 clinical trials, anywhere from several hundred to several thousand patients are tested in multiple sites to evaluate the overall benefit-risk relationship of the drug and then extrapolate the results to the general population. This phase alone can take three to four years and can cost between $50 million and $500 million.

Regulatory approval is the final phase before a new drug can go to market. In the US regulatory approval comes from the FDA. However, the FDA needs more people to ensure the regulatory process is done well.

"Studies indicate that out of every 5,000 to 10,000 screened compounds, only 250 enter pre-clinical testing, five enter clinical testing, and one is approved by the Food and Drug Administration. Only three in 10 of the drugs approved and marketed produce revenues to recoup their research and development costs."[3]

Improving Drug Development /

Most pharmaceutical companies are developing new models to improve the efficiency of the drug-discovery process. The following are some of the key steps being taken to improve efficiency:

- Reduce the drug-cycle time by decreasing costs, being more efficient and effective, and increasing the number of patents.
- Decrease the lag time for the stages during drug approval. To save costs it's important to find out early that a drug isn't going to succeed.
- Improve the utilization of people, teams, buildings, and campuses to support some standardization and reinforce an integrated global strategy.
- Outsource whatever is possible to improve response time and align core competencies.
- Overall focus on productivity, efficiency, and enhanced utilization.

Advances in computer technology and the new sciences—such as genomics, proteomics, and bioinformatics—will make it possible to identify new drug candidates at much earlier stages in the process, potentially saving a significant amount of time and money as well as providing better healthcare sooner. This will reduce research and development cycle times and costs. **With the cost of drug discovery escalating and the number of drug approvals declining over the past five years, pharmaceutical companies are re-evaluating their processes and developing adaptive clinical trials.** Flexible protocols are evolving and modifications are being made in clinical trials without undermining the validity and integrity of the trials. In layman terms this means that in order to achieve success at a faster rate the research team will look at the data after each patient then determine how to modify the study for the next patient. The process is more focused on continuous learning and on evaluation during each phase rather than at the end of a phase.

"By altering the original protocol to optimize the size and scope of confirmatory trials, little time is wasted between phases, often referred to as 'white space'."[4] Sometimes there is six months to a year between completion of Phase 2 and the start up of Phase 3. Time, money, and possible opportunities are lost by not having a more streamlined process. The industry is finding more success by adapting and thinking outside the box.

Pharmaceutical companies are beginning to share their data more freely on the Internet. This is beneficial for others who may be doing similar studies so that they can re-evaluate their work and possibly save time and money, or take their research in a different direction. One problem with sharing data on the Internet is that the data may be misinterpreted, creating possible concerns and arguments. **The fundamental concept of sharing data off the Internet is a very good one and should be encouraged.** GlaxoSmithKline (GSK) and other companies should be commended for doing this. The FDA should consider this as a requirement.

Another change developing in industry is adaptive clinical-trial designs to potentially reduce the size, length, and cost of clinical trials. The traditional Phase 1 and 2 stages of development may be replaced by a "learn phase" where the research team can "determine a compound's safety, efficacy, appropriate dosage range, and medical need, all in the context of investment costs."[5] After a compound has attained success in the learn phase it will be followed by a "confirm phase." The confirm phase is intended to replace the traditional Phase 2.

Another trend that is helping to improve how information is collected is Electronic Data Capture (EDC) and Electronic Health Records (EHR). The healthcare industry is also adopting electronic patient diaries, which are a more effective way for patients to report data and for industry to collect and study the data. Recent studies show that 90 percent of patients are compliant with the study protocols using electronic diaries and enter their data in real time.

Factors that could best prevent future delays in the US:

- EDC technologies
- Fewer intermediaries
- More money for patient recruitment
- Standardized Case Report Forms (CRFs)
- Site involved earlier in the protocol process

03.7 / A NEW WAVE OF PIPELINE GROWTH: PROJECTS WORLDWIDE (DISCOVERY-CLINICAL, 1993-2007)

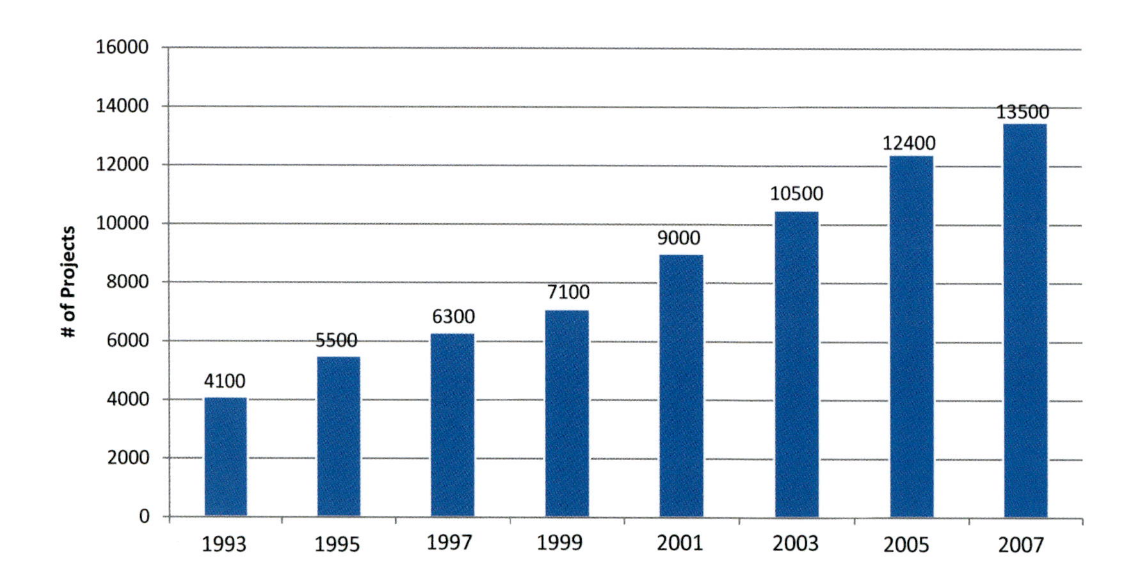

Time should be saved by optimizing the clinical trials with more efficient and effective data. These changes, which are starting to occur in the pharmaceutical industry, will likely require the research-and-development teams to adjust their standard operating procedures to facilitate mid-trial modifications. The adaptive-design process will be more reliant on the latest technology in order to minimize risk, and the latest computer technology can be used to simulate guided clinical trials.

The FDA's Critical Path Initiative promotes innovative trial designs to create a more effective model. Adaptive clinical-trial designs are being encouraged under this initiative, and this will result in better treatments being delivered at a faster rate to the patients who need them. The benefits of innovative trial design are beginning to become integrated across the industry.

"In a time when the drug development industry is under enormous pressure to reduce the cost of researching new drugs, the prospect of conducting shorter trials with fewer patients and the opportunity to save invaluable development time will eventually bring about mass adoption of innovative trial designs," explained professor Donald Berry at the University of Texas M.D. Anderson Cancer Center.

The chart above illustrates that the number of drugs in the pipeline have more than tripled from 1993 to 2007. The hit rate for successful drugs needs to increase, but unfortunately this number has decreased over the past five years.

US Food and Drug Administration /

Drug approvals by the FDA have dropped from 35 in 1999 to only 17 in 2007. The drug discoveries are called molecular entities, which are active ingredients that were never marketed in the US. The FDA designated the majority of new drugs as having "limited or no clinical improvement" over existing drugs.

This is a disturbing trend considering that more money is being invested in research, more collaboration is occurring between the various stakeholders, and larger amounts of significant data are being analyzed by the latest computer

Country	Regulatory Body	Clinical Application
Australia	THERAPEUTIC GOODS ADMINISTRATION (TGA)	CLINICAL TRIAL NOTIFICATION (CTN) / CLINICAL TRIAL EXEMPTION (CTX)
Canada	HEALTHCANADA	CLINICAL TRIAL APPLICATION (CTA) / DEMANDES D'ESSAIS CLINIQUES (DES)
China	STATE FOOD AND DRUG ADMINISTRATION (SFDA)	CLINICAL TRIAL APPLICATION
European Union	EUROPEAN MEDICINES AGENCY (EMEA)	CLINICAL TRIAL APPLICATION (CTA)
India	CENTRAL DRUGS STANDARD CONTROL ORGANIZATION (CDSCO)	INVESTIGATIONAL NEW DRUG APPLICATION (IND)
Japan	MINISTRY OF HEALTH, LABOUR AND WELFARE (MHLW)	CLINICAL TRIAL NOTIFICATION (CTN)
Russia	MINISTRY OF HEALTH	CLINICAL TRIAL APPLICATION
Singapore	HEALTH SCIENCES AUTHORITY (HAS)	CLINICAL TRIAL CERTIFICATION (CTC)
South Africa	REGULATORY AUTHORITY (DEPARTMENT OF HEALTH)	CLINICAL TRIAL APPLICATION
South Korea	KOREA FOOD AND DRUG ADMINISTRATION (KFDA)	INVESTIGATIONAL NEW DRUG APPLICATION (IND)
Switzerland	SWISSMEDIC	CLINICAL TRIAL NOTIFICATION (CTN)
United States	FOOD AND DRUG ADMINISTRATION (FDA)	INVESTIGATIONAL NEW DRUG APPLICATION (IND)

technology. "Industry analysts say the FDA is more cautious after drawing criticism in recent years for approving some drugs whose risks were found to outweigh their benefits after they went to market. Examples include painkillers Vioxx and Bextra. Both are now off the market."[6] The FDA approved only 64 percent of the applications they received from 1997 to 2005.

The chart above is a list of some international regulatory bodies. Successful international collaboration is another step toward simplifying and streamlining drug approval between countries.

The cost of research is the main reason why new drugs are so expensive. Pharmaceutical companies are concerned about profit loss due to generic drugs. The chart that follows illustrates that since 2002 the sale of generic prescription drugs have outsold patented prescription drugs. As this trend is expected to continue, it will put more pressure on pharmaceutical companies to reduce costs and to be more effective in how they conduct research.

Two news articles explained the impact generic drugs have had on two large pharmaceutical companies, Novartis and Wyeth. *The Shanghai Daily* reported in 2007 that Novartis had shares dip on sales of Lotrel by 75 percent, of Lamisil by 71 percent, and of Trileptal by 67 percent due to the availability of the equivalent generic drugs.[7] Slowing sales resulted in Novartis cutting 2500 jobs over the following two years. *The Bloomberg News* stated that Wyeth faces the possibility of losing 40 percent of its profit to generic competitors.[8]

The chart on the next page represents the amount of dollars spent within the USA and worldwide on generic drugs.

By 2013 most of the leading pharmaceutical companies will have blockbuster drugs whose patents will have expired. Historically, from two to six generic drugs compete for market share following the expiration of a patent on a blockbuster drug. However, due to the increased sales of generic drugs in recent years, more companies than ever before are competing against the original drugs.

03.8 / GENERIC DRUGS COST FROM 30 TO 80 PERCENT LESS THAN BRAND NAME COUNTERPARTS. MOST PEOPLE FEEL JUST AS SAFE TAKING A GENERIC DRUG AS A BRAND-NAME DRUG PARTLY BECAUSE THE GENERIC DRUG MUST ALSO RECEIVE FDA APPROVAL.

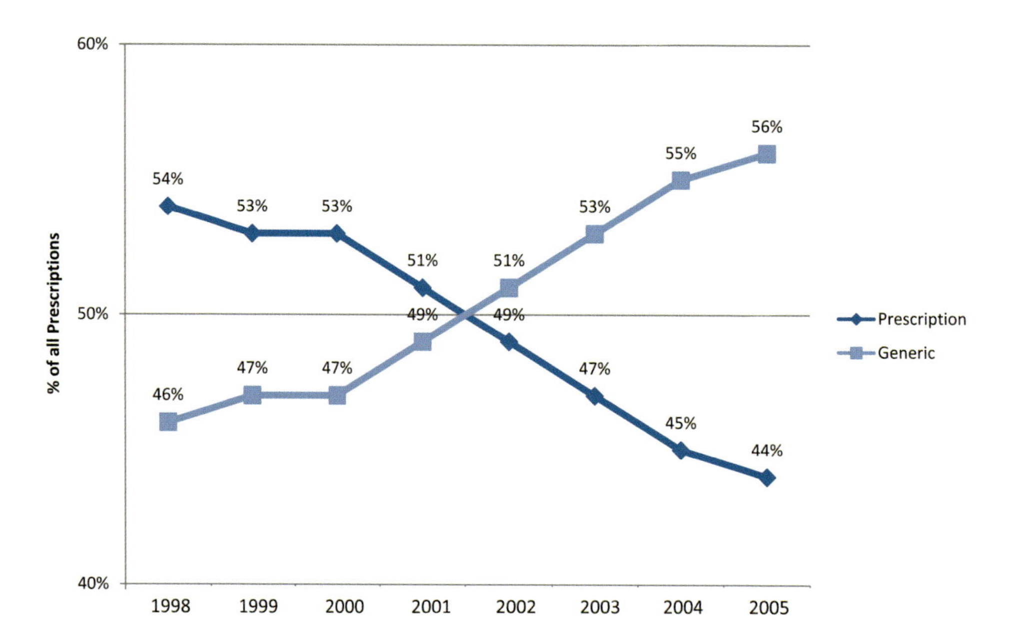

One advantage of generic drugs is that they help to reduce healthcare costs for consumers. Some pharmaceutical companies are now finding ways to make profits by focusing on generic drugs. The cost of healthcare cannot continue to escalate in the US. As all countries move forward, it is important to understand and to support entrepreneurship. The promise of reasonable profits will encourage private companies to undertake more productive research that in turn will result in the discovery of more affordable drugs.

The World's 10 Largest Pharmaceutical Companies

Company	Country	2008 Employees	Total Revenue (Million US$)
PFIZER (WITH WYETH)	USA	137,127	$71,130
JOHNSON & JOHNSON	USA	119,200	$61,095
GLAXOSMITHKLINE	UK	103,483	$45,447
BAYER	GERMANY	108,600	$44,664
HOFFMAN-LA ROCHE	SWITZERLAND	78,604	$40,315
SANOFI-AVENTIS	FRANCE	99,495	$39,997
NOVARTIS	SWITZERLAND	98,200	$39,800
ASTRAZENECA	UK/SWEDEN	67,400	$29,559
ABBOTT LABORATORIES	USA	68,697	$29,527
MERCK & CO.	USA	74,372	$23,850

US Leads Pharmaceutical and Biotech Industries /

The drug industry consistently reports profit margins approaching 30 percent of revenue, and spends slightly more than 20 percent of its revenue on research and development. Among the world's 50 top pharmaceutical companies, 22 are US owned, 14 are European, 10 are Japanese, three are Australian and one is Middle Eastern. None of the top 50 are from China or India.

It's interesting to compare the research and development expenditure (see the end of Chapter 2) by country to the revenues generated by big pharmaceutical companies (in US$). Most pharmaceutical companies invest approximately 25 percent of their revenues into research.

ANNUAL RESEARCH EXPENDITURE

USA	$303,340,000,000
Japan	$144,237,000,000
Pfizer	**$71,130,000,000**
Germany	$68,799,000,000
Johnson & Johnson	**$61,095,000,000**
France	$46,262,000,000
GlaxoSmithKline	**$45,447,000,000**
Bayer	**$44,664,000,000**
Hoffman-La Roche	**$40,315,000,000**
Sanofi-Aventis	**$39,800,000,000**

The World's 10 Largest Biotech Companies

Company	Country	Employees (2006)
AMGEN	USA	20,100
GENENTECH	USA	10,533
GENZYME	USA	9,000+
UCB	BELGIUM	8,477
GILEAD SCIENCES	USA	7,575
SERONO	SWITZERLAND	4,775
BIOGEN IDEC	USA	3,750
CSL	AUSTRALIA	2,895
CEPHALON	USA	2,515
MEDIMMUNE	USA	2,359

Of the world's top 100 Biotech companies, 75 are owned by US companies, and the top seven US biotech companies generate more revenue than the other 93 companies combined.

Academia /

The USA's Top Fundraising Universities (By US$ millions received in 2009)

1.	STANFORD	640
2.	HARVARD	601
3.	CORNELL	446
4.	PENNSYLVANIA	439
5.	JOHNS HOPKINS	433
6.	COLUMBIA	413
7.	SOUTHERN CALIFORNIA	368
8.	YALE	358
9.	UCLA	351
10.	WISCONSIN, MADISON	341

To stimulate the transfer of technology from universities to businesses, many governments—especially the Australian, Canadian, and US—have encouraged universities to patent their inventions. Indeed, academia is increasingly the birthplace of marketable ideas. Northwestern University received $700 million in 2007 for selling partial rights to future royalties from a new blockbuster drug, Lyrica, used to treat pain related to fibromyalgia, shingles, and diabetes.[9] Earlier that year, New York University received $650 million for its rights to partial royalties on a drug called Remicade used to treat inflammatory diseases. If sales exceed benchmarks, the university will get a larger percentage of the revenue.[10] Professors receive a majority of their funding for research from grants, then from endowments, and finally from student tuition.

The following is a summary of the drug discovery process as it evolves from academia:

1. An academic scientist designs an experiment to answer an important question.
2. The scientist applies to the government to fund the research.
3. The money pays for students and fellows who conduct the research.
4. The results are published in journals, which advance the field.
5. An invention may result. This may lead to a patent, which then is licensed to a start-up company.
6. With a monopoly granted by the patent, the company attracts venture capital. If it is successful, the company grows.
7. Years later, the discovery becomes a therapy for patients.

Types of Research /

Research and development covers a broad range of scientific and business activity. Research is systematic study directed toward fuller scientific knowledge or understanding of the subject. It is classified as either basic or applied according to the objective of the sponsoring agency. Development is the systematic application of knowledge.

Basic Research /

The goal of basic research is simply to advance scientific knowledge. It is systematic study directed toward fuller understanding of the fundamental aspects of phenomena and of observable facts. There is no specific application or immediate commercial investment in mind, although the research may lead to financial benefits.

Research institutions influence both directly and indirectly the type of research and development conducted by their scientists and engineers. The most direct influence is the decision to fund specific research and development projects. This influence tends to be weaker in academia than in industry or government because academic researchers are freer to seek outside research and development funding. This relative autonomy, cushioned by the tenure system, makes universities and colleges well suited to carrying out basic research (particularly undirected basic research).

Federally supported basic research, aimed at understanding many features of nature—from the size of the universe to the nature of subatomic particles, from the chemical reactions that support a living cell to interactions that sustain ecosystems—has been an essential feature of American life and has helped to drive economic success for more than 50 years. While the outcomes of specific projects are never predictable, basic research has been a reliable source of new knowledge. This in turn has fueled important developments in fields ranging from telecommunications to medicine, and has yielded positive rates of economic return and has created entirely new industries that have provided high-tech, high-wage jobs.

Academia believes in "open architecture," meaning that the knowledge produced by research should be made public to encourage innovation. Universities thrive on the free flow of information, which allows researchers to quickly build on the work of others, typically even before it is published. On the other hand, private industry contributes little to basic research. However, amplified basic research can benefit companies by boosting human capital (through attracting and retaining academically motivated scientists and engineers) and by strengthening innovative capacity (the ability to absorb external scientific and technical knowledge). Industries that invest the most in basic research are those whose new products are most directly tied to recent advances in science, such as the pharmaceutical industry.

Applied Research /

Applied research is the application of systematic study to gain the knowledge required to meet a specific, recognized need. In private industry, applied research includes discovering scientific knowledge, the application of which has specific market value with respect to products, processes, or services.

The business sector spends more than three times as much on applied research than it does on basic research, and provides about half of the applied research funding in the USA. The level of applied research in an industry reflects both the market demand for substantially new and improved goods and services, as well as the level of effort required to transition from basic research to technically and economically feasible concepts.

Development /

Development is the systematic application of knowledge gained from research directed toward the production of useful materials, devices, systems or methods, including design, development, and improvement of prototypes and new processes.

Since the late 1990s, a new informal innovation system has evolved in the US based on close collaborations and increasing alliances among industry players, universities, and government labs. Much of the work has been centered on entrepreneurial needs, but in the future some research needs to be balanced into the bottom line to support people in need even if the investment can't be justified financially.

"Higher prices are supposed to spur research for lifesaving medicines ... the fact (is) that most drug companies spend far more on advertising than on research, more on research for lifestyle drugs than for disease-related drugs, and almost none on research for the diseases prevalent in the poorest countries, such as malaria or schistosomiasis," writes Joseph Stiglitz in *Making Globalization Work.*[11]

Research needs to be more strategic and needs to consider the broader goals of society. **The United Nations reports that less than 10 percent of healthcare research is spent on the problems affecting 90 percent of the world's population**. More incentives are needed to encourage drug discoveries that can help those who are most in need.

It is also important for the public to understand that most good things often have some negative side effects. Research must address any potential negative issues before new products are released to the public; the benefits must significantly outweigh the problems. It is the responsibility of good science to minimize mistakes, but at the same time society must understand that it is unlikely that new developments will be 100 percent flawless.

International Research Model /

Many companies have research and development facilities around the world. In his book *The Post-American World*, Fareed Zakaria discusses the happy face (☺) where the smile represents development of a product from conception to sale. "At the top of the curve one starts with the idea and high-level industrial design—how the product will look and work. Lower down on the curve comes the detailed engineering plan. At the bottom of the U is the actual manufacturing, assembly, and shipping. The rising up on the right of the curve are distribution, marketing, retail sales, service contracts, and sales of parts and accessories. In almost all of the manufacturing, China takes care of the bottom curve and America the top—the two ends of the U where the money is."[12]

Many countries are collaborating with companies to help grow their local economies. Today, for instance, a molecule can be discovered in Germany, refined in Japan, clinically trialed in many different countries at the same time, manufactured in India or Ireland, then distributed worldwide.

The key is clarity of focus, insight, and experience in research and product development. Confidence and intelligent risk-taking are also a key part of the equation that will usually equal success. Companies will have a much better chance of success if they have products that are widely distributed in the global market. ■

04. SUSTAINABLE SOLUTIONS

Sustainable Research /

Much research is focused on sustaining ecological support systems. As a global society, we now understand the need to become energy efficient and to reduce waste. At the same time, private industry sees increasing opportunities for creating solutions and increasing profits. **Opportunities for making new discoveries and for making money over the next 10 to 20 years may be the highest in sustainable research.**

The World Commission on Environment and Development defines sustainable development as "meeting the needs of the present generation without compromising the ability of future generations to meet their needs." This definition echoes the Native American Iroquois Confederacy philosophy of the Seventh Generation that advises that all decisions should be made with consideration of the impact they may have on descendants seven generations to come.

No matter what the terminology—green building, high-performance building, sustainable design, or environmental design—the goal is to create a superior built environment for little or no extra cost. In green building, operational expenditures are lower over the life of buildings while occupant health, safety, and productivity are improved, and liability is reduced.

It is proposed that the projects move beyond simply sustaining current conditions to a model that is regenerative in nature. Establishing more aggressive approaches to key sustainability indicators such as carbon neutrality, water-use reduction, and a commitment to reducing persistent toxicity, will serve as a basis for a regenerative systemic model. By thinking in terms of regenerative design, it is possible to not only reduce the negative impact of construction and operations in terms of resource, energy and water use, but to create a positive impact through the integration of current and future environments, and ultimately, living cities and campuses.

04.2

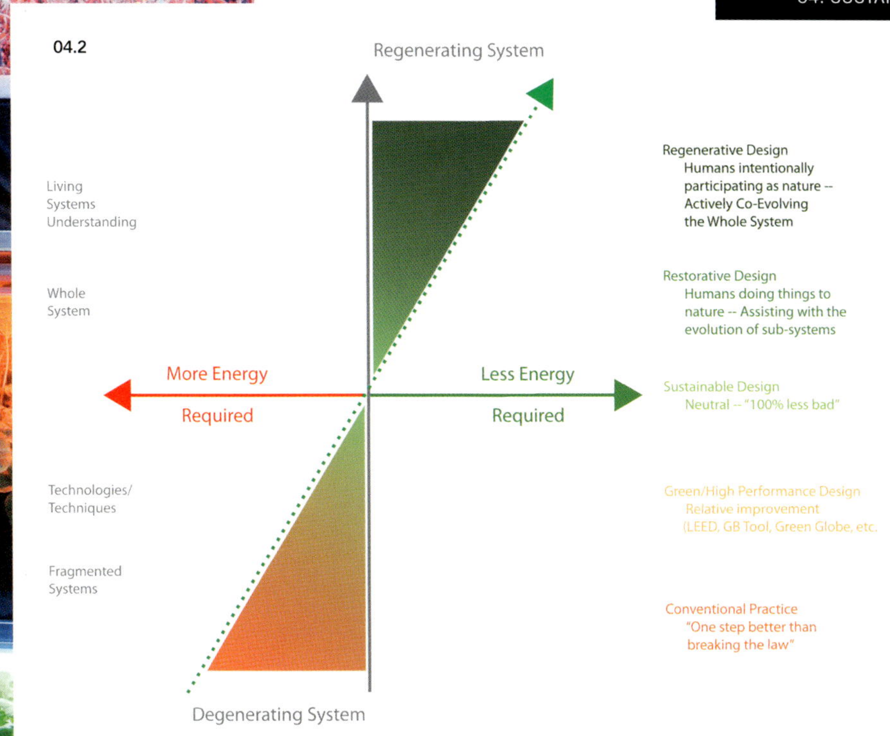

Global Impact /

The natural environment provides us, free of charge, all we need for survival. Ecosystems purify the air we breathe and the water we drink and convert waste into resources. Biodiversity provides food and medicine, all the while maintaining genetic variety to ward off pests and diseases. The environment has the ability to sustain plant, animal, and human life. A healthy environment is everyone's challenge. In rich countries, the by-products of industry and agribusiness poison the soil and waterways. In developing countries massive deforestation, harmful farming practices, and uncontrolled urbanization are commonplace. The 20 percent of the world's population living in the industrialized countries consumes nearly 60 percent of the world's energy, but the developing world's share is rising rapidly.

We now have indisputable evidence that the earth's climate is changing. This is largely due to the burning of fossil fuels such as coal and oil to run our vehicles, homes, and factories. It is also due to mass deforestation—each day at least 80,000 acres of forest disappear, and deforestation is estimated to be responsible for about 20 percent of global carbon emissions.

"It's becoming clear that humans have caused most of the past century's warming by releasing heat-trapping gases as we power our modern lives. Called greenhouse gases, their levels are higher now than in the last 650,000 years," reported *National Geographic*.[1]

According to the National Oceanic and Atmospheric Administration and to the National Aeronautics and Space Administration (NASA), over the past 100 years the Earth's average surface temperature has increased by about 1.2 to 1.4 °F. And according to the Intergovernmental Panel on Climate Change (IPCC), 11 of the 12 hottest years since thermometer readings became available occurred between 1995 and 2006. Rising temperatures are linked to natural disasters. As a result of global warming, glaciers are melting, sea levels are rising, cloud forests are disappearing, and countless animal species are becoming extinct.

04.1 / VEGETABLES ARE GROWN UNDER LIGHTS POWERED BY SOLAR ENERGY AT A LABORATORY IN BEIJING ON DECEMBER 16, 2009.

04.2 / SYSTEMS THINKING CAN RESULT IN RADICALLY EFFECTIVE INTEGRATED DESIGNS THAT OPTIMIZE THE CONSUMPTION OF ENERGY, WATER, AND OTHER RESOURCES.

As for the ramifications of global warming, we only now see the tip of the shrinking iceberg. As sea levels rise, shorelines erode and salt water intrudes into freshwater aquifers, threatening drinking water sources and crop production. As *Time* magazine reported in December 2007, "Droughts are baking the US southwest, Australia, and sub-Saharan Africa; floods are devastating Bangladesh; and Central America is reeling from powerful hurricanes. Not all of these events can be tied absolutely to global warming, but all of them will surely become more frequent and intense as the world warms—ultimately threatening the lives and livelihoods of hundreds of millions of people."[2]

Deaths	Year	Event
UP TO 3.7 M	1931	YELLOW RIVER FLOOD, CHINA
UP TO 900,000	1938	YELLOW RIVER FLOOD, CHINA
UP TO 830,000	1956	FLOOD IN SHAANXI PROVINCE, CHINA
UP TO 500,000	1970	BHOLA CYCLONE, BANGLADESH
UP TO 287,000	2004	INDIAN OCEAN EARTHQUAKE, TSUNAMI

Research is needed to better understand the changes that are occurring and to provide solutions to minimize natural disasters and to slow the destruction of the planet.

Sustainable Leadership /

In February 2008, 480 business leaders and institutional investors from around the world met at a United Nations (UN) summit in New York and pledged to invest US$10 billion over two years in technologies to reduce greenhouse-gas emissions. The group also called on the US Securities and Exchange Commission to pressure companies into disclosing their potential impact on climate change, and it called on the US government to enact laws to reduce greenhouse-gas emissions by up to 90 percent by 2050.

These investors see climate change as having the potential to be economically devastating while at the same time presenting great economic opportunities. Mindy Lubber, president of Ceres Investor Coalition, told *The Guardian* news agency: "This action plan reflects the many investment opportunities that exist today to dent global warming pollution, build profits and benefit the global economy. Leveraging the vast energy efficiency opportunities at home and abroad holds especially great promise for investors."[3]

Carbon Neutral /

When we have carbon in our environment, we have more heat. This heat is stored in the oceans, creating an increase in the water temperature. The rise in temperature increases the frequency and the amount of damage created by weather-related occurrences. Research over the past 100 years has tracked these changes against the increased use of petroleum.

Carbon emissions are generated in approximately three areas—buildings, transportation, and industry—each of which contributes to about one-third of the emissions.

Architects try to focus on several key points:

1. There can be a 10 to 15 percent reduction in energy needs in the USA from people simply changing their habits. Turning off lights and appliances when not in use, installing temperature controls in rooms, making greater use of natural ventilation, driving fuel-efficient cars, and making use of public transportation options are all simple solutions that can make a considerable difference.

2. When possible, buildings can be orientated so they are facing north/south to take full advantage of the sun's energy.

3. Buildings can be constructed using more effective insulation, better entry vestibules, airtight windows and doors, and rain screens to protect the exterior facades.

4. Approximately 19 percent of the energy used in a typical research building is a result of lighting. By making better use of natural light, LED light fixtures, light sensors, and lower ambient light requirements, energy usage for lighting can be cut by at least half.

5. In most countries, natural ventilation is a building-code requirement, however, building codes in the USA fall behind in this respect. Not only can buildings be effectively cooled using natural ventilation, but room tolerances need to increase in the winter and summer months—65 °F (compared to 68 °F) in the winter, and 76 °F in the summer (compared to 70–72 °F).

6. Develop strategies to make use of the thermal mass, or heat, held within concrete walls and slabs—heat is released from the walls and ceilings at night helping to heat the building during the colder months, while in the warmer months the concrete draws in the heat to help cool the building.

7. Renewable energy technologies have attracted a great deal of attention in recent years as a potential solution to the sustainability challenges associated with the construction and operation of buildings. Unfortunately, however, at the present stage of development the vast majority of renewable-energy systems are not economically viable for commercial-scale projects. The cost of photovoltaic panels, for instance, has been cut in half over the past five years but the payback is still more than 30 years. The case for wind turbines is similar. Recently, a large number of public and private organizations have committed a tremendous amount of money to the research and development of renewable energy technologies. These massive expenditures have led many in the industry to believe that the cost performance of renewable technology systems, photovoltaics in particular, will improve dramatically over the next five years, becoming cost-competitive with conventional energy systems. Solutions for renewable energy will be the most significant research discoveries in the first quarter of the 21st century.

National Perspective /

The US government owns approximately 445,000 buildings with a total area of 3 billion square feet. It leases an additional 57,000 buildings representing a total area of 374 million square feet. The Federal Leadership in High Performance and Sustainable Buildings Memorandum of Understanding works to green all government facilities, including those that are either under construction or are to be constructed.

The newly enacted Energy Independence and Security Act of 2009 addresses a number of issues related to building and infrastructure projects:

- A budget of $3.75 billion over five years has been allocated for the Department of Energy's (DOE) Weatherization Assistance Program.
- Section 422 establishes a commercial-buildings' zero-energy initiative to achieve a national goal of zero-net-energy use in buildings built after 2025.
- Retrofit all pre-2025 buildings to zero-net-energy use by 2050.
- Section 431 commits the government to the 2030 Challenge for all Federal Buildings, an incremental reduction in building-energy use from 55 percent below 2003 levels by 2010, and energy neutral by 2030.
- Improved integration of solar-power plants into regional electricity transmission systems.
- Section 656 directs DOE to establish a cost-shared Renewable Energy Innovation Manufacturing Partnership Program to make awards to support research and development on advanced manufacturing processes, material, and infrastructure for renewable energy technologies. Further goals are to increase domestic renewable energy production and better coordinate federal, state, and private resources through partnerships. Solar, wind, biomass, geothermal, energy storage, and fuel-cell systems are eligible forms of equipment.
- Carbon capture and sequestration research, development, and demonstration. Carbon Capture and Storage (CCS) is a term used to describe a combined set of chemical and geological processes involved in the capture of carbon dioxide, which is emitted during the combustion of a conventional fuel, and the long-term storage of that carbon dioxide in a stable underground geologic formation. Many view this system of technologies as a means of reducing our carbon emissions as we await the transition from carbon-based energy resources to carbonless alternatives. However, the viability of these costly and complex CCS systems as a mainstream solution to our carbon-emissions problem has yet to be proven. Indeed, there has yet to be a fully functional demonstration of a large-scale integrated CCS system. So far only the independent implementation of the various subsystems have been seen to work successfully.

Also included in the act are provisions to support the development of low-emitting and renewable technologies abroad (for more information on this topic go to http://energy.senate.gov/public/_files/RL342941.pdf):

- The promotion of clean and energy-efficient energy technologies in foreign countries. Regardless of where renewable energy technology breakthroughs are made, be it in countries of the developed or developing world, it is essential to the long-term survivability of the human race that the information be disseminated across national borders with immediacy and fluidity. Climate change and the widespread decline of environmental support systems are systemic global problems requiring systemic global solutions. If and when a truly game-changing breakthrough is made, we must realize that we no longer have the luxury of being able to wait for its piecemeal implementation.
- International Clean Energy Foundation to fund emission-reduction projects abroad.
- Green jobs!

The Carbon Disclosure Project (CDP) is an independent not-for-profit organization aiming to create a lasting relationship between shareholders looking for value and corporations looking for commercial opportunities presented by climate change. The CDP's goal is to facilitate a dialogue, supported by quality information, from which a rational response to climate change will emerge. The CDP provides a coordinating secretariat for institutional investors with combined assets of over $57 trillion under management. On their behalf the CDP seeks information on the business risks and opportunities presented by climate-change and greenhouse-gas-emissions data from the world's largest companies. As of 2008 the CDP represented

3,000 investors. Currently more than 60 percent of the companies listed on the S&P 500 are represented (https://www.cdproject.net). The business world is beginning to recognize that climate change represents a real risk to future growth and profits, and it's likely that a nation-wide carbon market will be created in the next few years.

State Perspective /

California recently proposed a program that would allow the government to charge an annual fee to any business that emits greenhouse gases. While this proposal has not yet been passed into law, it signifies what may be coming down the pike. The proposal is one of the first to recognize that there are real costs associated with emitting carbon dioxide and other greenhouse gases.

Twenty-four states have implemented renewable energy portfolio standards to mandate renewable-energy production as part of the utility fuel mix. Some states, represented by a block of states in the northeast and another on the west coast, have developed trading platforms similar to the Kyoto Protocol to help commoditize carbon emission and generate business wealth from emission reductions.

In August 2007, the leaders of six western states and two Canadian provinces agreed to their own regional climate pact, aiming to cut greenhouse-gas emissions to 15 percent below 2005 levels by 2020.

City Perspective /

We have long heard the adage "think globally, act locally." Cities present a unique opportunity to address major environmental and social challenges; they manage local energy and water systems, trash collection, air quality, waste removal, and transportation plans. For example, the US Mayors Climate Protection Agreement, introduced by former Seattle mayor Greg Nickels, has been signed by nearly 700 mayors from across the US. The agreement aims to set emission-reduction targets based on the Kyoto Protocol, despite the absence of the US as a signatory.

With a visionary plan to address today's challenges, cities can effectively make moves to improve their operations, and ultimately save money. A plan needs to be in place to create targets and metrics in key areas of stewardship such as waste reduction and energy efficiency.

Once the plan is in place, a necessary key to success is sufficient stakeholder buy-in. This does not mean that department heads simply mandate improvements, but that every employee in the organization is given the opportunity to offer his or her ideas and experience to help guide the plan to its stated goals. But in order for this to succeed, everyone in the organization needs to understand what the goals are and how they, as individuals, can help to achieve them. It's important that organizations make a commitment to educating staff.

Using an integrated design process, cities can analyze their impact on the environment in a holistic fashion and find ways to optimize benefits across the board. In 1989, New York City was facing an estimated $8 billion cost for the construction of adequate water-filtration plants to provide 1.4 billion gallons of high-quality drinking water to 9 million people each day. Rethinking conventional models for water delivery, the city devised an innovative plan to protect the Croton and Catskill/Delaware watershed systems, which together span 1,969 square miles and are located 200 miles from the city.

Working with the roughly 500 farms and 60 towns in the watershed, NYC proposed an integrated and long-term watershed protection program that targeted upstream water to improve quality down stream. Through the success of these actions the US Environmental Protection Agency (EPA) waived the filtration requirement and NYC was able to lay claim to the first such upstream/downstream collaboration to promote water quality while preserving the economic objectives of the farming economy in the upstream watershed. With an estimated cost of only $507 million, the program provided massive savings compared to filtration and had the added benefit of long-term protection of the watershed ecosystem.[4]

Yet another display of integrated-solution development can be found in Los Angeles (LA) where city officials were able to realize comparable savings when faced with a similar infrastructure problem. Like many urban environments, LA has a large amount of paved surface area that is impervious to rainwater and restricts the amount of water that is able to percolate into the underground aquifers. A $10 billion proposal was in place to expand the stormwater infrastructure to meet the capacity needed. Instead, a local non-profit organization responded with an alternative plan that was ultimately tested in demonstration installations and adopted. For a fraction of the cost, LA was able to instead invest in on-site stormwater management strategies that yielded a much greater payback.[5] Greensburg, Kansas is the first city in the US to require LEED Platinum certification and a 42 percent reduction in energy use for all new municipal buildings.[6] In years to come, it can be expected that this level of commitment will increase among cities, as well as on campuses where serious consideration will need to be made regarding campus buildings and infrastructure (see the following section). Pragmatically, an Integrated Design Process is used to achieve this level of integration and holistic design, whether it involves water-management plans or building design.

It is important to remember that sustainability is a process and not an end goal. Many processes are in place to reduce impact, but the key is to institutionalize such processes to create a truly sustainable organization. The end goal, then, is one of continuous improvement in which new objectives and measures are routinely established to improve performance. As such, sustainability is an iterative process in which opportunities are assessed, implemented, and measured continuously.

University Campus Perspective /

More than 400 university presidents from across the country have signed onto the American College & University Presidents' Climate Commitment. While standards are currently being set for carbon reductions, this initiative has demonstrated the strong commitment of universities across the country.

When thinking in terms of campuses, many opportunities are exposed for integrating systems between buildings and campus infrastructures. Stormwater management and grey/black water-treatment and reuse not only conserve water and promote aquifer recharge, but also provide landscape features and biodiversity by providing a host of ecologies to the campus. Many of these systems can be used as educational tools for chemistry, civil engineering, biological sciences, and design.

There is a trend, especially in academic environments, to design entire sustainable communities. **An educational environment offers invaluable opportunities for collaboration among university leaders, faculty, students, and the architectural and engineering design teams. This should be a hands-on productive experience for the students, who, as the next generation of decision makers, can advance these ideas. Furthermore, faculty will have the opportunity to test and challenge research ideas on site for further development.** Given our current environmental challenges, all campuses should have a master design plan for sustainability that is incorporated into a regional strategy. Such a plan will allow for changes in utility systems, electrical supply, and stormwater management to take place over 20 to 25 years.

Universities such as Arizona State, Columbia, Harvard, Stanford, and Wisconsin have created programs for sustainability to help the global environment. Universities educate most of the people who develop and manage society's institutions. For this reason, universities bear profound responsibility to increase the awareness, knowledge, technologies, and tools to create an environmentally sustainable future.

Reducing Risk /

Performance Contracting /

To help finance their commitment to reducing energy use, campuses and cities are turning to performance contracting. Performance contracting relies on Energy Service Companies (ESCO) to analyze and identify potential savings from energy, operational, and infrastructure improvements. The upgrades are paid for by the savings realized from the improvements. "According to the US Department of Energy, between 1990 and 2003 at least $15 billion was invested in 1,300 alternatively financed projects (including PCs) at federal facilities, municipal governments, universities, schools, and hospitals."[7]

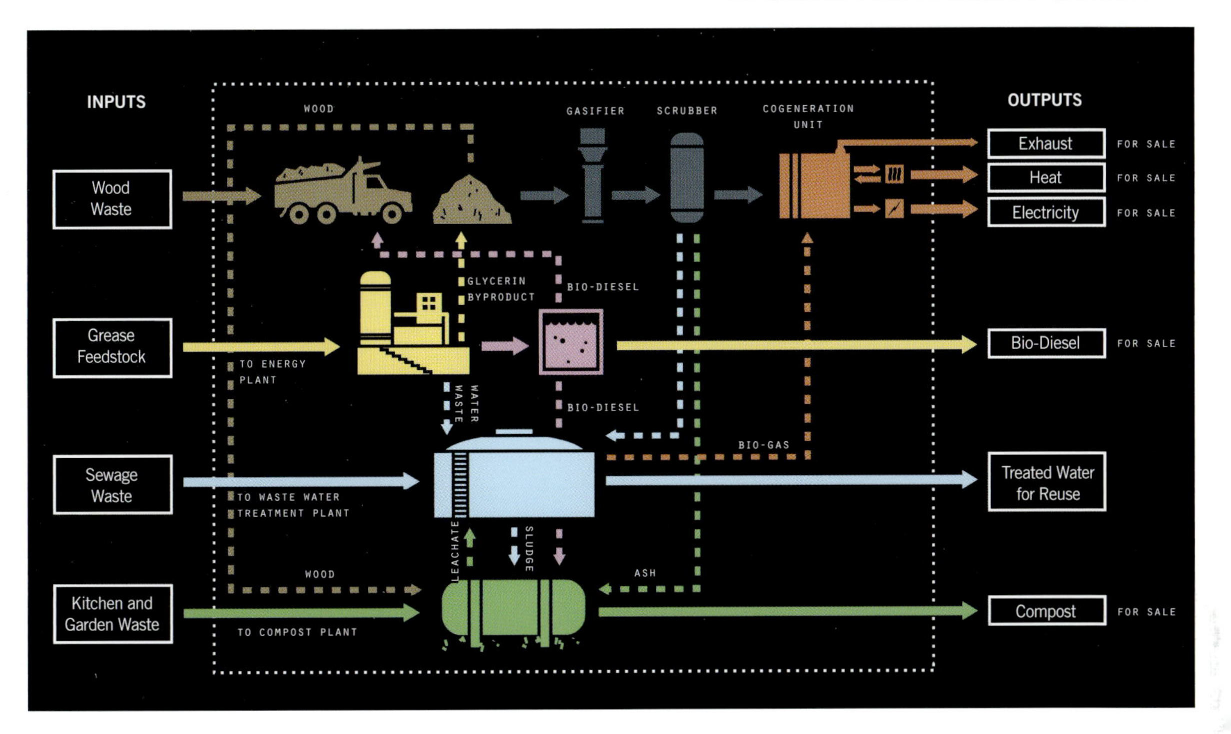

04.3 / AN EXAMPLE OF A MASTER DESIGN PLAN FOR SUSTAINABILITY.

The ESCO can guarantee savings up to a specified amount. Typically, the guarantee will be lower than the estimated savings so that operational savings will provide a positive cash flow once the financing is paid each period.

In the B-O-O model, these energy generating systems are Built, Owned and Operated by a contractor who will contract with the building owner to sell the power at a fixed rate over a defined period of time. The owner does not purchase, own, or operate the installed equipment. This methodology insulates a building owner from innovative and rapidly developing technology that may be outdated in a short period of time. The owner is also insulated from costs related to energy production, efficiency of components, or other expenses.

Insurance /

Capital and Life Cycle Costs /

The insurance industry is responding to marketing research data indicating that there are reduced risks with some aspects of buildings that are designed to be green, and this has resulted in a reduction of premiums of up to 10 percent. Transformation of the market is evident as the insurance industry recognizes risk mitigation and extended liability related to sustainable design practices and construction. Companies such as Fireman's Fund Insurance and Lexington Insurance Company provide specific policies for green buildings, and more policies are being offered by more companies all the time.

Many insurers are providing policies that, in the event of the loss of an existing building, will cover the cost of constructing a green-certified building. Certified green buildings may have components that are not currently covered by traditional insurance polices such as vegetated roofs, and alternative power and energy systems. However, policies should recognize the superior features of green buildings by including a rate credit. To qualify for some policies, the building must be certified green by the US Green Building Council (USGBC).

Commissioning coverage applies to green certified as well as to traditionally constructed buildings. Following the loss of a building, policies may cover the cost of hiring a commissioning engineer to ensure that the new building systems such as heating, ventilating and air conditioning (HVAC), electric, and plumbing operate at peak performance and in alignment with one another.

Reducing Waste /

Buildings and Systems /

The 2007 USGBC Research Funding Report stated that: "Buildings also have a significant impact on human health. Indoor air typically contains between two and five—and occasionally greater than 100—times more pollutants than outdoor air. As a result, poor indoor air quality in buildings has been linked to significant health problems such as cancers, asthma, Legionnaires' disease and hypersensitivity pneumonitis."[8]

Other important benefits from the research into more sustainable environments are the discoveries that are made that can improve human intelligence. An example of this is found in a study conducted by the Centers for Disease Control and Prevention (CDC).

The study showed reducing lead levels in the blood by 10 ug/dl raised children's IQ an average of 2.6 points. Prior to 1976, lead was used as an additive in gasoline to improve engine performance. In response to a growing body of research indicating lead's negative impact on human health, in 1978 the substance was banned from use in gasoline and other products such as paint.

The CDC studies showed an immediate decline in blood lead levels—over a 25-year period, there was a 15 ug/dl drop, and today only a slight trace of lead is found in the blood of small children. Over this same period of time IQs have increased by 4 percent, which is significant. The EPA estimated the value of one IQ point in 1998 at $8,350, and in 2007 four points were worth approximately $42,000. It is clear that the quality of our environment impacts us in many ways, which makes it even more important to improve the quality of all lives.

Building industries—such as architecture, engineering, manufacturing, construction, and operations—employ over 1.7 million people and make up a significant part of the gross domestic product (GDP)—an estimated $1 trillion per year. According to the 2007 USGBC Research Funding Report, this represents the largest economic sector in the USA.

Building operations account for 40 percent of US energy use, and this number increases to an estimated 48 percent when the energy required to manufacture building materials and construct buildings are included in the figure. Building operations alone contribute more than 38 percent of the country's carbon-dioxide emissions and over 12 percent of its water consumption. And 35 percent of all non-industrial waste—136 million tons of landfill debris annually—is a result of the building industry.

Building design, construction, and operations represent what may be the most promising solution set. Buildings consume 36 percent of total energy and are responsible for 65 percent of electricity consumption in the US, resulting in 30 percent of greenhouse-gas emissions. In addition, a considerable amount of our outdated building stock is replaced annually, representing an additional opportunity for improvement. According to Architecture 2030, in the next 23 years 1.75 billion square feet of building will be demolished, 5 billion square feet will be newly constructed, and an additional 5 billion square feet will be remodeled or retrofitted.

Initiated by architect Edward Mazria, the 2030 Challenge seeks to leverage the massive amount of building construction as a viable strategy to combat carbon emissions. By setting aggressive goals for energy efficiency, the 2030 Challenge has spurred innovation and dialogue around the issue of building emissions, and it has been endorsed by the American Institute of Architects (AIA), the American Society of Heating, Refrigerating and Air-Conditioning Engineers (ASHRAE), and the USGBC. The 2030 Challenge calls for immediate energy reduction of 50 percent for all new projects, 60 percent by 2010 and an additional 10 percent every five years, culminating in climate neutral building operations by 2030.

A summary on Carbon Reduction Strategies (CRS), completed by the Chicago chapter of the AIA, identified architectural strategies that can impact the carbon emissions of projects for immediate through to long-term reductions, and at various cost scales. For example, strategies that can achieve both reduced carbon emissions and cost savings include turning off lights and idle equipment, installing programmable thermostats for night-

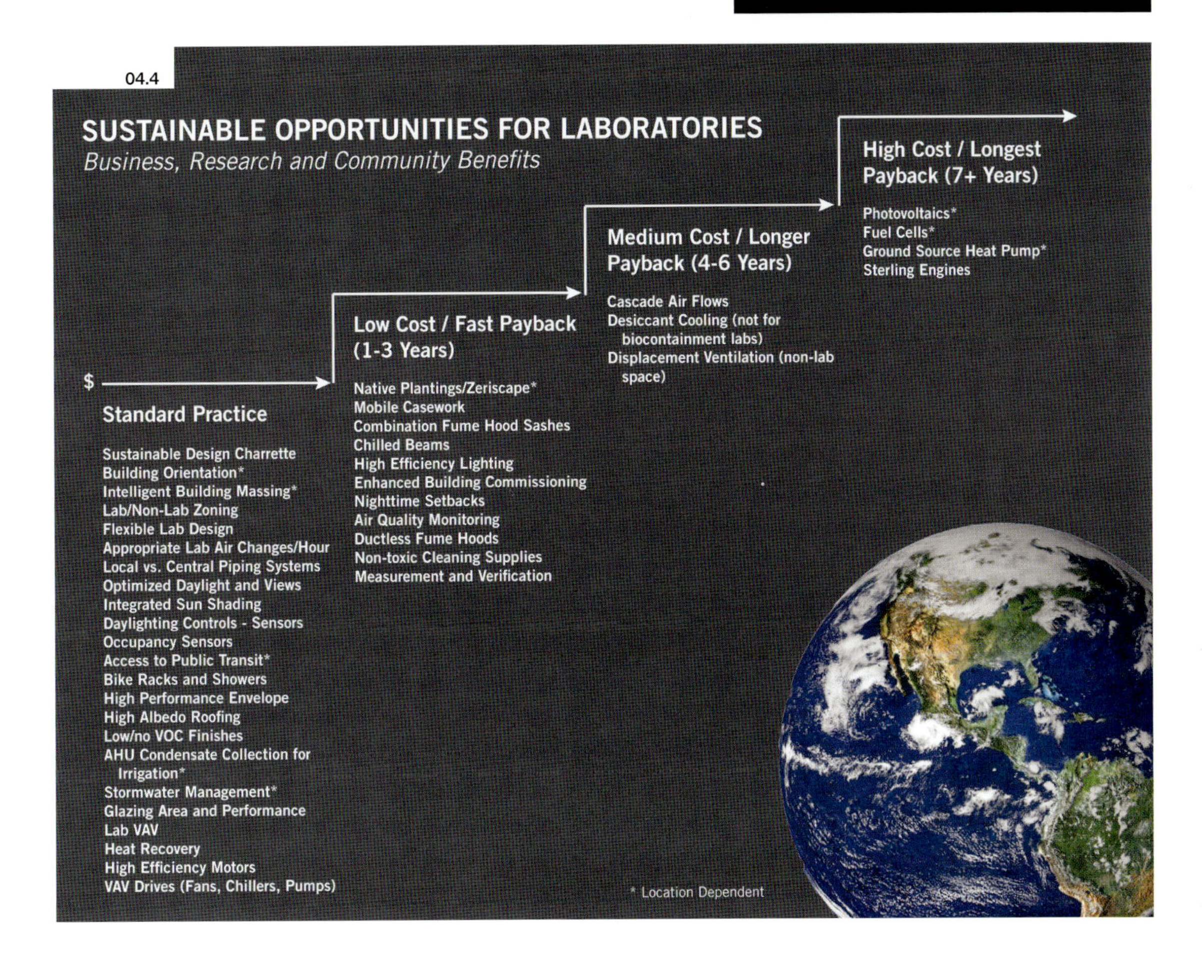

time setbacks when facilities are not being used, regular cleaning of the cooling coils in the mechanical system, and servicing of the economizer cycle. According to the DOE, an improperly serviced economizer cycle can add as much as 50 percent to a building's annual energy bill. The CRS outlines other strategies with a range in carbon reduction timeframes and costs.

The costs associated with the life cycle of buildings are frequently overlooked and unaccounted for. This occurs in many organizations, such as universities, because the capital budget is funded separately from the operations budget. This means that there is no incentive for even a modest increase in upfront costs despite a considerable payback in operations and maintenance. First costs account for less than 10 percent of money spent on a facility over its lifetime, whereas from 60 to 85 percent of the cost associated with the building is spent over time through operations and maintenance. It is possible to significantly reduce and recover life-cycle costs within three to five years, even with high-value features.

The illustration above lists many possible solutions when designing laboratory facilities.

Improving Quality of Life /

In building their high-rise corporate headquarters, a large commercial client wanted not only to build a Leadership in Energy and Environment Design Gold certified building, but also to make an outstanding indoor environment that enhanced productivity. Intuitively they felt that providing a sky garden, visible from each floor, would offer a place for stress relief and an environment that encouraged collaboration. It was hypothesized that views to two sky gardens should be provided on each floor for a maximum benefit to all employees, regardless of where their desk was located. This option would eliminate valuable leasable floor space, so the client wanted to see scientific evidence of the actual benefits before committing to the plan.

For performance indicators the research team found that natural settings create positive market identity, which could mean increased business. Natural settings also increase staff satisfaction, thus positively impacting recruitment and retention of top employees. One of the reasons for this effect was that views onto nature evoked the most popular hobby among workers, which is gardening. Other productivity benefits included an improvement in innovation and creative problem solving and high-order cognitive function. **Research confirmed the supposition that vegetated areas of respite facilitate communication and interaction, which in turn builds trust among staff.** In terms of quantitative productivity gains, views of vegetation restore cognitive performance following fatiguing challenges.

04.5 / THE SINGAPORE CREATE PROJECT FOCUSED ON INTEGRATING GARDENS FROM THE EXISTING SITE UP INTO THE ATRIA AND THEN ON TOP OF THE ROOFS.

For psychological indicators much of the research revolved around stress and positive mental states. Gardens provide a positive escape to recuperate from stress. Stress reduction occurs because plants absorb, diffract, and/or reflect background noise. This decreased stress allows creativity and innovative thinking to flourish. Attentiveness is also improved. Another interesting finding is that plants create positive perceptions. Reception-area visitors reported that they thought the spaces were more ornate, interesting, cheerful, expensive, tidier, and quieter.

For physiological indicators, plants create ideal humidity levels, which result in fewer colds and flu viruses being transmitted throughout the building. Plants also reduce symptoms that contribute to absenteeism such as headaches and eye and skin irritations.

Better Research Buildings /

Among building types, research facilities are particularly important candidates for green building techniques—they use five times as much energy and water per square foot as a typical office building, yet are seldom the focus of energy conservation campaigns within those organizations. One of the main causes for such a high premium on the operational costs of chemically intensive research laboratories, as compared to more conventional work spaces, is the need to protect workers from the effects of hazardous materials. Risk is a function of hazard and exposure. Conventional approaches to minimizing risk in laboratory spaces attempt to reduce levels of human exposure while maintaining a high level of hazard within the space. A new more effective approach, both in terms of cost and occupant health and safety, would be to address the intrinsic hazard being encountered in the space to eliminate the need to control exposure.

Laboratories are energy-intensive for many reasons:

- For safety and research-related reasons they house several containment and exhaust devices that facilitate adherence to intensive ventilation requirements, including "once through" air. The fact that they require "once through" air is a proximate cause. The root cause of this problem is the use of volatile hazardous materials with which contact must be strictly prevented.
- They house a great deal of heat-generating equipment.
- Scientists often require 24-hour access to the facilities.
- Irreplaceable experiments require fail-safe redundant backup systems and uninterrupted power supply or emergency power.

Through working together, architects, engineers, building owners and researchers are making great strides in decreasing the environmental impact of research facilities. Designing from a holistic perspective reveals significant opportunities to improve energy efficiency and reduce or eliminate harmful substances and waste, while meeting or exceeding health and safety standards—especially important in research laboratories. It is easily possible today not only to dramatically reduce energy consumption, but also to select ecologically acceptable materials without compromising architecture or comfort. But this requires holistic, interdisciplinary thinking, which has to prevail from the first planning ideas right through to the final decision-making process.

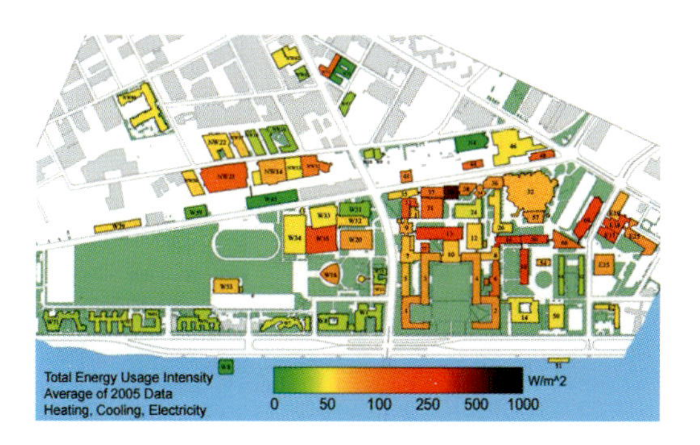

04.6 / AT MIT CAMPUSES IT IS CLEAR WHERE THE RESEARCH BUILDINGS ARE LOCATED—IN THE DARKER COLOR BLOCKS. THIS IS TRUE FOR RESEARCH CAMPUSES AROUND THE WORLD WHERE THESE BUILDINGS REQUIRE MORE ENERGY THAN ANY OTHER TYPE OF BUILDING.

Many strategies, such as using products with recycled content, are being applied to all building types. But for research facilities, the greatest impact comes with energy and water savings, accumulated during the operation and maintenance phase of the building's life cycle, including:

- Reducing the amount of outside air used for ventilation.
- Insulating hot water, steam, and chilled water piping.
- Keeping condenser water as cool as possible, but not lower than 20 °F above chilled water supply temperature.
- Employing a heat-recovery system.
- Installing an economizer at the boiler. (The water-side economizer will help with humidity controls.)
- Maintaining hot water for washing hands at 105 °F. A good practice is to use local hot water tanks at kitchens, restrooms, and other areas instead of using a central hot water supply.
- Using ultra-low-flow toilets (0.5 gallons per flush) and automated controls such as infrared sensors for faucets.
- Harvesting rainwater and reusing "gray water" from sinks for irrigation.
- Metering to verify the efficient operation of the building's systems.
- Testing mechanical equipment for performance at least once a year.

Whole-building systems-analysis software simulates the operation of the entire HVAC system. Furthermore, commissioning, which typically provides financial payback within one or two years, ensures that systems are operating at top efficiency. Many new laboratories also include direct digital-control energy management systems to assist in running the building more efficiently to save more money by using less energy.

It is now possible, and becoming more common, to design more efficient buildings with little if any additional cost. As mentioned above, the first objective is to reduce waste and to reduce the amount of energy required to run laboratories. The second objective is to make buildings energy self-sufficient. The third objective is to use these buildings as research tools to benchmark sustainable solutions and develop new ideas.

Another key principle of sustainable design is maximizing the availability of natural daylight. In addition to reducing energy use, natural daylight improves employee productivity by creating a more comfortable and a more pleasant environment. Wherever possible, natural light should be the primary source of illumination; artificial lighting should be thought of as only a supplement to natural light. Designers should strive to have natural light in most laboratory and public spaces so that, from almost anywhere in the building, people can look outside, see what the weather is like, and orient themselves to the time of day.

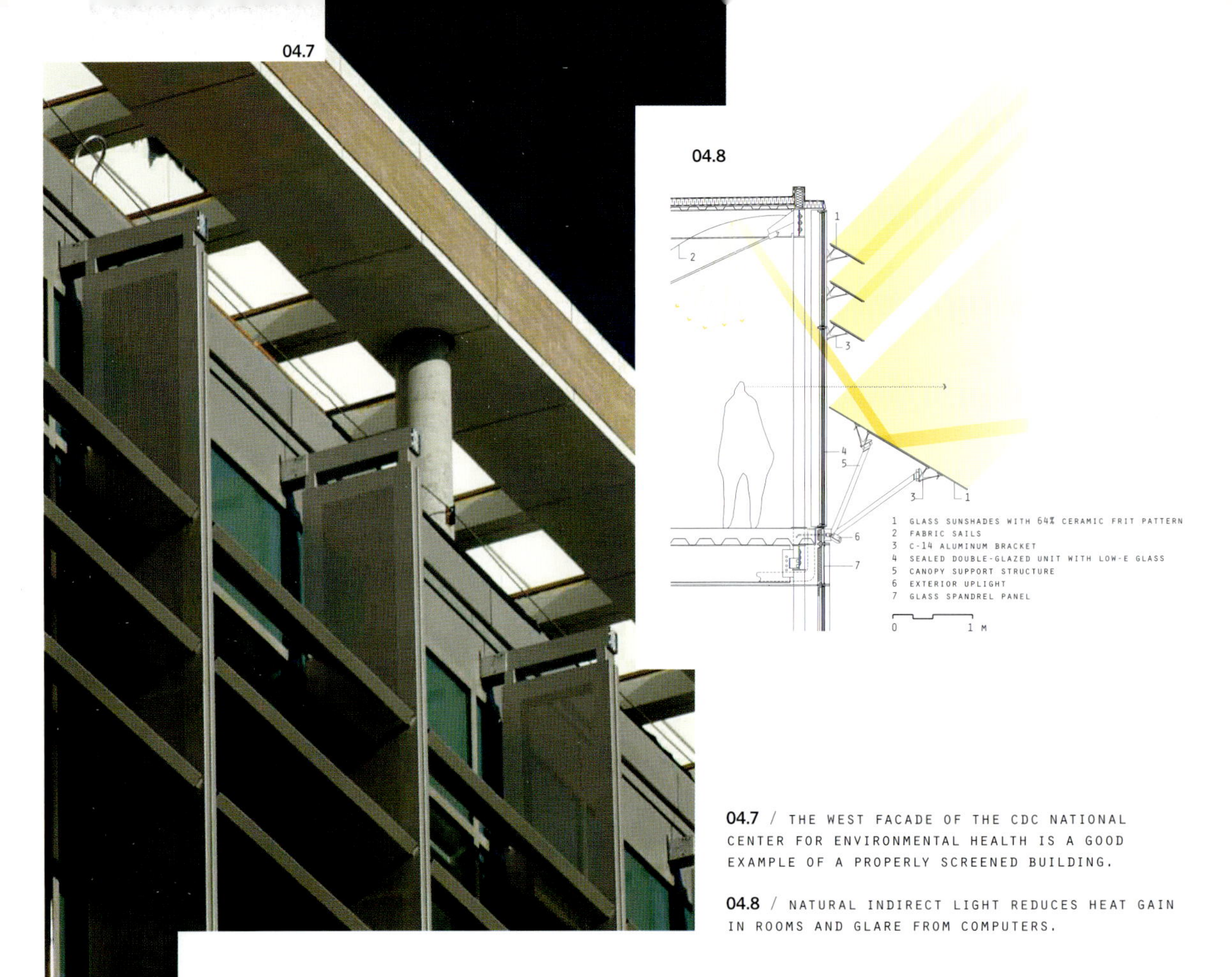

04.7 / THE WEST FACADE OF THE CDC NATIONAL CENTER FOR ENVIRONMENTAL HEALTH IS A GOOD EXAMPLE OF A PROPERLY SCREENED BUILDING.

04.8 / NATURAL INDIRECT LIGHT REDUCES HEAT GAIN IN ROOMS AND GLARE FROM COMPUTERS.

Indirect light minimizes both glare and heat gain from the sun. Typically, the first 20 feet of depth at the perimeter of buildings can be entirely lit by natural light during the day. The use of light shelves, which block direct light and provide for natural indirect light, can extend the natural-light zone in buildings up to 30 feet. Clerestory windows and skylights can be used to draw even more natural light into buildings.

Green Chemistry /

Green chemistry—the practice of designing chemicals that are environmentally benign yet commercially viable—is coming into its own. And, as always, advances in science beget advances in business. The principles of green chemistry are being adopted in a growing range of industries, such as the biomedical industry, the electronics industry, as well as other consumer goods industries.

"Green chemistry has already turned maize into biodegradable plastics, developed non-toxic solvents and dramatically reduced the toxic byproducts from the manufacture of popular pharmaceuticals like ibuprofen. It is vital to the production of interior finishes in Toyota's new electric cars, made in part from kenaf, an annual grass plant," reported Soyatech, provider to the soybean and oilseed industries.

Green chemistry reduces pollution at its source by designing products and processes that reduce or eliminate the use and generation of hazardous substances. Practitioners of green chemistry attempt to eliminate intrinsic chemical hazards by thoughtful design at the molecular level. In doing so, they must evaluate, in a holistic way, the agent's eventual impact on human health as well as on a wide variety of other physical and biological systems in the natural environment. Greening a process reduces its complexity, the energy input required, and the volume and toxicity of waste and substance produced.

Questions green chemists need to ask are:

- Is the material or its by-products toxic?
- How will this affect those working with it?
- Is it renewable?
- Why use a large molecular compound when only a small part of it is required for a specific process?

In order to answer many of these questions, green chemists have developed a number of new performance metrics that go beyond traditional measures of reaction efficiency to determine the aggregate environmental impact of chemical products and processes. By incorporating such things as energy consumption, the complexity of reaction conditions, and the hazard levels and volume of waste produced, green chemists are able top develop a comprehensive picture of the net effects of a chemical product or process on human health and the environment.

Though Europe and Asia have long advocated green-chemistry agendas as a least-cost method of achieving many environmental goals, it has only recently gained traction in the US. The Pollution Prevention Act of 1990, which established a national policy to prevent or reduce pollution at its source, provided the funding and institutional support required to give the go-ahead to green chemistry research and development. The EPA, charged with enforcing the act, continues to promote green chemistry through research grants, public-private partnerships, and its prized Presidential Green Chemistry Challenge Awards.

Even though green chemistry programs have been supported for 20 years, the world is now seeing the results of this research. **Public knowledge of green chemistry today is about where sustainable design was 10 years ago, however, hopefully it will take less than 10 years for green chemistry to support long-term solutions in the marketplace.**

According to the EPA, some of the technologies nominated for the Presidential Green Chemistry Challenge Awards include:

- Eliminating the use or generation of 1.2 billion pounds of hazardous chemicals and solvents each year—enough to fill over 5,000 railroad tank cars or a train over 62 miles long.
- Saving over 16 billion gallons of water each year—enough to supply a city the size of Baltimore, Maryland.
- Eliminating 57 million pounds of carbon dioxide releases to air each year—equal to taking nearly 37,000 automobiles off the road.

04.9 / GREEN CHEMISTRY IS A NEW APPROACH TO THE DESIGN OF CHEMICAL PRODUCTS AND PROCESSES WHICH, AT ITS CORE, SEEKS TO ELIMINATE THE GENERATION AND USE OF HAZARDOUS SUBSTANCES. GREEN CHEMISTS REJECT THE ASSUMPTION THAT FOR CHEMICALS TO PROVIDE COMMERCIALLY VIABLE FUNCTIONALITY THEY MUST NECESSARILY BE ASSOCIATED WITH ADVERSE HEALTH AND ENVIRONMENTAL CONSEQUENCES.

04.10

Headlines declaring the dangers of harmful chemicals used in toys, cough syrup, toothpaste, shampoo, and other everyday products have given rise to the green chemistry movement. And in addition to these concerns, experts say that employing the principles of green chemistry has the potential to positively impact on everything from climate change to global water and food supplies, and to reduce the risk of industrial disasters such as the devastating gas leak in 1984 at the Union Carbide pesticide plant in Bhopal, India.

04.10 / ELECTRON MICROGRAPH OF NANOPARTICLES ON THE SURFACE OF AN ALUMINUM-BASED CATALYST COMMONLY USED IN LABORATORY SYNTHESES. SOME OF THE MOST SIGNIFICANT ADVANCES IN THE FIELD OF GREEN CHEMISTRY HAVE BEEN RELATED TO THE DEVELOPMENT OF REUSABLE CATALYTIC SUBSTANCES THAT DRIVE CHEMICAL REACTIONS WITHOUT THEMSELVES BEING CONSUMED. THROUGH THE USE OF THESE NEW CATALYTIC PROCESSES, GREEN CHEMISTS HAVE BEEN ABLE TO DRIVE DOWN PRODUCTION COSTS, REDUCE REACTION TIMES, DECREASE ENERGY CONSUMPTION, AND IMPROVE MATERIALS-USAGE EFFICIENCY.

A stern 2006 report by the California Research & Policy Center stressed the need for the widespread practice of green chemistry: "**Every day, the US produces or imports 42 billion pounds of chemicals, 90 percent of which are created using oil, a non-renewable feedstock. Converted to gallons of water, this volume is the equivalent of 623,000 gasoline tanker trucks (each carrying 8,000 gallons), which would reach from San Francisco to Washington, D.C., and back if placed end-to-end. In the course of a year, this line would circle the earth 86 times at the equator.** These chemicals are put to use in innumerable processes and products, and at some point in their life cycle many of them come in contact with people—in the workplace, in homes, and through air, water, food, and waste streams. Eventually, in one form or another, nearly all of them enter the earth's finite ecosystems."[9]

Industry is increasingly seeing green chemistry's inherent advantages. Berkeley Cue, Jr., who started the green-chemistry initiative at pharmaceutical giant Pfizer, told CNET News.com: "Chemical processes that reduce waste translate into huge savings."[10] Preventing costly pollution problems has become a business necessity; manufacturers are spending hundreds of millions of dollars to clean up dioxins, perchlorate, mercury, and asbestos. Examples of environmental destruction caused by these harmful chemical agents and the associated costs with remedying their effects are numerous:

- In 2004, DuPont agreed to pay up to $600 million for environmental damage caused by the production of Teflon and Gore-Tex.
- General Electric is spending tens of millions of dollars to clean up PCBs it discharged into the Hudson River.

- In Bohpal, India, more than 15,000 people were killed and 150,000 to 600,000 were injured in 1984 when 40 tons of methyl isocyanate (MIC) leaked from a pesticide plant owned by Union Carbide (now owned by Dow Chemical Co.).
- A 2005 benzene spill in China contaminated the water supply of millions of people in the Jilin Province.

Today, however, the list of successes in the business applications of green chemistry is growing rapidly. "DuPont's Teflon production pollution problem was solved by rethinking how the molecules making up Teflon are put together. It now uses carbon dioxide as a surfactant rather than the toxic Perfluorooctanoic acid," reported Soyatech. "To produce one of its most popular drugs, the pharmaceutical giant Pfizer revised a complex four-step process that produced toxic wastes into a one-step process using ethanol, saving millions of dollars. Other big pharma companies have made similar changes in their manufacturing processes, saved millions of dollars, and now regularly win environmental awards from the US Environmental Protection Agency."[11] There is a recent trend to focus on green chemistry at the beginning of the drug-development process in order to fully capitalize on the economic and environmental benefits.

Other success stories include the use of supercritical fluid chromatography (SFC) in drug discovery and development. Traditionally, chromatography—a process that is used for the analysis of separation of reaction byproducts—involves large amounts of volatile solvents that can create waste-disposal and recycling problems. SFC is not only faster than traditional chromatography, but also uses recyclable CO2 instead of petrochemical-derived hydrocarbons as the solvent, resulting in a much greener and ultimately lower cost operation.

Merck scientists, are using high-throughput screening equipment and new micro-scale analytical methods that allow them to perform chemistry on a smaller scale in all aspects of testing and development. Like SFC, these technologies are speeding up development and minimizing waste, byproducts, and excess uses of solvents and reagents. Coupled with the increased use of catalytic methodologies, this has led to a greening of the entire process—from drug discovery through development.

04.11 / THIS SCANNING TUNNELING ELECTRON MICROSCOPE AT THE BROOKHAVEN NATIONAL LABORATORY IN UPTON, NEW YORK IS JUST ONE OF A SUITE OF PRECISION INSTRUMENTS THAT PROBE THE BEHAVIOR OF CHEMICAL SUBSTANCES AT THE SUB-NANOMETER SCALE ALLOWING GREEN CHEMISTS TO DESIGN MOLECULES WITH IMPROVED FUNCTIONALITY AND REDUCED TOXICITY.

04.12

Virtually all commercial chemicals are currently derived from petroleum feedstocks. Green chemists, however, see enormous opportunities in not only being able to mimic natural chemical processes but also in using natural materials as renewable feedstocks. This new mindset requires that chemists be retrained in their thinking and in the way they approach conventional problems. Green chemistry proponents also speak of the urgency to teach green practices to scientists in developing countries.

Green chemistry and toxicology are increasingly being taught in both introductory and advanced science courses, requiring students to forge new methods and to give consideration to the effects of their work on human health and the natural environment. To mirror this new curricular focus many large academic institutions are beginning to incorporate the principles of green chemistry in the programming and design of new laboratory facilities. Labs in which green chemistry is practiced have been found to use substantially fewer hazardous materials and produce significantly less toxic waste. As a result these labs use from 75 to 80 percent less energy than traditional labs resulting in tremendous energy savings and reductions in their operating costs.

Reversing our damaging impact on the environment requires long-term commitment to research, but it must be done. Paul Anastas, professor of chemical engineering at Yale University and assistant director for the EPA, stated in the fall of 2008 that **during the 20th century we were trying to do all the right things, but doing them the wrong way. The 21st century focus is do the right things the right way. Green chemistry will be a key component to accomplishing this goal.**

Renewable Energies /

Photovoltaic systems, wind-energy harvesting, geothermal energy, and biomass-electrical generation usually carry a high initial return on investments. A main advantage of renewable-energy technology is that the majority of the costs are upfront, while the "fuel" cost is essentially free. Adoption of renewable-energy technology to meet energy needs in rural areas offers an opportunity to leapfrog the traditional development paradigm characterized by centralized fossil-fuel power plants. Power provided by renewable-energy technology can be appropriately scaled to meet the demands of the market.

Sun–Photovoltaics /

In the US, the federal government and many state and local governments are providing incentives for renewable technologies, such as photovoltaics (PVs), to enable them to become more viable in the market. Currently, PVs are priced at roughly $8 per watt installed. The payback depends on a number of variables such as climatic conditions and available solar resources, and also the local electrical utility rate. For instance, in some parts of California where electricity costs remain high, many incentives are provided by the state and local utilities, and as there is generally a high level of solar income, a three-to-four-year payback on photovoltaics is possible. In other areas, though, without the incentive structure or the sunshine, the payback may remain at 12 years or more.

Retrofits make up 98 percent of the PV market, meaning that they are simply added to existing buildings. Building-integrated photovoltaics (BIPVs) are electricity-generating materials that are used in place of conventional building materials in parts of the building envelope such as the roof, skylights, canopies, windows, or facades. As the production of PVs continues to escalate and technologies for BIPVs improve, the cost should decrease because BIPVs eliminate the marginal cost of installation. When integrated into glazing, for instance, the installation cost is absorbed by the cost that would have been incurred to install glazing without PVs.

BIPVs can be incorporated into new construction as well as retrofitted into existing structures. The main advantage of integrated PVs over more common non-integrated systems is that the initial cost can be offset by reducing the amount spent on building materials and labor that would normally be used to construct the part of the building that the BIPV modules replace. Since BIPVs are an integral part of a building's design, they generally blend in better and are more aesthetically appealing than other solar options.

BIPVs are capable of offsetting a good portion of a building's electrical requirements. For example, at the University of Wisconsin-Green Bay campus, BIPV vision glass was substituted for traditional windows in the wintergarden of Mary Ann Cofrin Hall. This system covers 2,000 square feet and generates about 12,500 kWh annually. Another example is the Frederick C. Murphy Federal Center in Waltham, Massachusetts, which houses irreplaceable federal documents. Its failing EPDM roof was replaced with a system that uses flat thin film, flexible, amorphous silicon panels integrated with a Sarnafil EnergySmart roof system. This saves the General Services Administration (GSA), the government agency responsible for the construction of many public buildings, about $67,000 per year in electrical costs.

04.12 / GREEN CHEMISTRY LABORATORIES REQUIRE FEWER FUME HOODS AND ARE MORE EFFICIENT TO RUN.

04.13

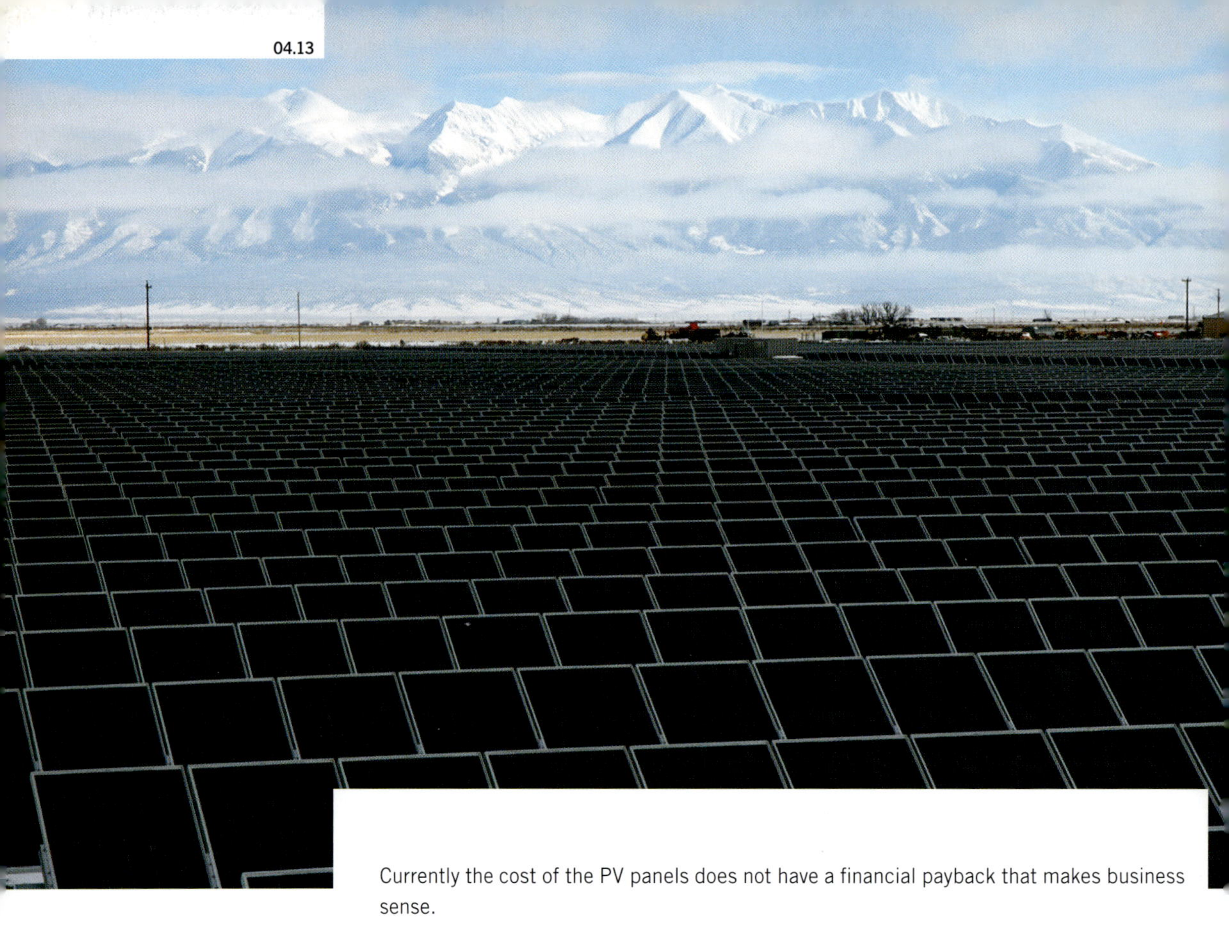

Currently the cost of the PV panels does not have a financial payback that makes business sense.

04.14

Many governmental financial incentives are available for the use of BIPV materials. The site www.dsireusa.org lists many state incentive programs. The federal government also offers an accelerated five-year depreciation and a first-year 10 percent investment tax credit.

Like many others, I firmly believe that within the next five years there will be significant improvements in the performance and cost efficiency of solar panels. When this happens we will see major shifts around the world in building renovations and the construction of new facilities. This will be especially important in developing countries that currently don't have a large infrastructure to work with so will be able to capitalize on this new technology.

A less tangible payback for the use of BIPVs is the prestige that many buildings obtain by being perceived as highly energy efficient, and the good public relations enjoyed by the owners for being responsible corporate citizens. The fact that BIPV materials are comparable in price to many of the traditional high-end materials, such as stone, gives added impetus for their use in place of these materials. As well, BIPVs blend seamlessly, and often unnoticed, into the aesthetics of the building.

Air–Wind Turbines /

Integrated wind turbines are simply buildings where wind turbines have been incorporated into the building design. This can take many forms, and turbines can be either architecturally integrated into the design or simply attached to the building structure. One exception to this is building-assisted wind turbines, where the physical building design

04.15

has been conceived in such a way as to actually funnel and direct wind in very specific pathways to increase wind flow to large turbines that are integrated into the building.

I am responsible for the design of the Oklahoma Medical Research Foundation's new research building. The building will be completed by the Fall of 2010. On the roof there are 24 wind turbines in the shape of a double helix.

Offsetting high energy costs is the fact that wind is free and inexhaustible. Once the capacity to capture energy from the wind has been acquired, the power itself is free, and will always be so. Most turbines are virtually maintenance free, especially the smaller varieties, which often have only two moving parts—the hub on which the blades and unit rotate, and the bearings on which the coils spin. Turbines are estimated to last from 20 to 30 years. Even so, high capital costs and lack of subsidies keep wind power relatively expensive for now, though several manufacturers estimate payback periods of five years or less. More research is needed to verify such claims. However, in rural areas, building-integrated wind can be very cost competitive, especially if coupled with solar power, compared to the costs of transmission and cabling over long distances.

In many countries wind and solar farms have been installed in large open areas, although wind farms are typically four to five times more effective than solar farms. Companies and institutions are purchasing large turbines and installing them in wind farms where the most electricity can be added to the grid. The local power company then credits the utility company servicing the building based on the amount of energy that has been generated by the wind. This is found to be the most cost-effective way to incorporate renewable energy into a building's construction. Many US states are creating wind farms and encouraging companies to partner with them to accelerate the effective growth of green renewable energy. These large renewable farms will support country, state, and regional grids, and this transformation is also beginning to take place around the world.

04.13 / IN AREAS WHERE THERE IS A LOT OF SUNSHINE, SOLAR ENERGY CAN PROVIDE A CLEAN AND INEXPENSIVE SOURCE OF POWER.

04.14 / AT BIOPOLIS, SOLAR PANELS ARE INTEGRATED INTO THE GLAZING TO HELP GENERATE ELECTRICITY, TO PROVIDE SHADE, AND TO SHOW OFF THE LATEST TECHNOLOGY.

04.15 / THIS FILM IS AN EXAMPLE OF A BIPV THAT IS INTEGRATED INTO THE BUILDING, IN MANY CASES BETWEEN TWO PANES OF GLASS.

Biofuels /

The use of fuel derived from organic material is expected to reduce greenhouse gases, improve vehicle performance, boost rural economies, and help protect ecosystems and soils.

Researchers worldwide are focused on the development of biofuels. Biofuel is any fuel derived from biomass. Biomass is organic material that has inherent chemical energy content and thus can be used to provide heat, make fuels, and generate electricity. Crops, cooking oil, and even methane fumes from landfills can be used as biomass to produce biofuel. Over the next 10 to 15 years, energy-efficient solutions are expected to increasingly drive worldwide economics—even surpassing the economic impact of technology in the 1990s.

The outlook for biofuels is not all rosy, however. The forecast does not guarantee blue skies and clean air. There is growing worry that biofuels may not be significantly cleaner in the long run than fossil fuels. Even if they are cleaner, there may not be enough land or water to sustain the millions of acres of crops required to produce the fuel. Because growing, harvesting, and processing crops to produce biofuels involves energy, much of which is now provided by fossil fuels, the process is not carbon-dioxide neutral. Most of the 124 ethanol plants in the US, for example, use natural gas and even coal to run the processors. And because ethanol corrodes existing pipelines, it must be trucked to where it's required.

Because of limitations and costs, biofuel production will most likely be a modest part of a portfolio of energy solutions. Global biofuel production tripled from 4.8 billion gallons in 2000 to about 16 billion in 2007, but still accounts for less than 3 percent of the global transportation fuel supply.[12] Even as biofuel production grows exponentially, world oil production is expected to grow 30 percent by 2030 and production of unconventional fossil fuels will increase even faster, according to the US Department of Energy.

04.17

OMRF

Anyway you look at it, the development of efficient biofuels promises to be a bumpy ride. Just like the stock market, one day the news is up, the next day it's down, and all the while fortunes (and lives) hang in the balance. The only way out of this thicket of uncertainty is a full-out commitment to research. Much of the biofuel research is focused on technological advances and efficiency gains—higher biomass yields per acre and more gallons of biofuel per ton of biomass. Researchers are seeking better ways to convert plants to ethanol and other fuels, exploring the potential of biomass (renewable resources) and studying the environmental and agricultural impact of producing biofuels on a grand scale.

One success story is Brazil, which produces 32 percent of global biofuel production in the world, and is the world's largest producer of sugar-based ethanol, accounting for around half of the world's total output. In 2005, Brazil produced about 3.52 billion gallons, most of which was used domestically to fuel automobiles, which typically run on a 25 percent ethanol gasoline blend. In 2006, after a three-decade-long alternative energy campaign, Brazil achieved energy independence. As of 2007, ethanol accounted for about 20 percent of the country's fuel supply.

Conclusion /

Sustainable research is, and will continue to be, a key area of focus by governments, private industry, and academia. It takes only three weeks to re-educate electricians on sustainable solutions. All tradespeople will need to be continually educated on the latest opportunities and technology. Within the next 10 years we can expect significant breakthroughs in research discoveries, not only because we must find solutions, but also because solutions will be financially feasible creating significant opportunities to make money (similar to the development of the dot-com technology in the 1990s).

The Future of Renewable Energy Research /

The following section is information I gathered from conversations and support from Otto Van Geet at the National Renewable Energy Laboratory (NREL). We are currently working together with his group and Labs 21; both are an excellent resources to collaborate with.

The world faces many energy challenges. Fossil fuels are a finite resource that will eventually be depleted. The burning of fossil fuels releases CO2 into the atmosphere, which is the major source in the rise of atmospheric carbon-dioxide levels (greenhouse gases) and a major contributor to global warming.

04.16 / THE OKLAHOMA MEDICAL RESEARCH FOUNDATION'S NEW BUILDING WILL BE SEEN FROM MANY AREAS IN OKLAHOMA CITY AND PROVIDES A STRONG NEW BRAND IMAGE FOR THIS NOT-FOR-PROFIT RESEARCH INSTITUTION.

04.17 / THE DOUBLE HELIX, REPRESENTING THE DNA MODEL, IS THE OKLAHOMA MEDICAL RESEARCH FOUNDATION'S LOGO. WIND TURBINES WILL GENERATE ELECTRICITY AND SUPPORT A NEW BRAND IMAGE OF INNOVATION FOR THE ORGANIZATION.

04.18

04.19

The only good way to dramatically decrease the release of CO2 is to reduce energy use and meet remaining energy needs with renewable energy. This requires accelerated research into renewable energy from scientific innovations to market-viable solutions. Research includes the understanding of the resources that can be used for renewable energy to the conversion of these resources to renewable forms of energy, and ultimately to their use in homes, commercial buildings, and vehicles. Renewable-energy research will follow the full research and development cycle—from basic scientific research through to applied research and engineering, testing, scale-up, and demonstration, all of which will require many new research facilities.

Here is a brief overview of some of the major areas of renewable research. (For more detailed information see www.nrel.gov.)

Wind /

Wind is one of the fastest growing and lowest cost sources of energy in the world. To increase the development of this energy source, research is required to develop larger, more efficient, utility-scale wind turbines for land-based and offshore installations, as well as more efficient, quieter small-wind turbines for distributed applications. Major research areas in wind-turbine design to identify innovative modifications that will decrease component weight and cost include turbine rotors, blades, and drivetrains.

Solar /

Solar technologies use the sun's energy to provide heat, light, hot water, electricity, and even cooling for homes, businesses, and industry. Major areas of research focus on PV systems, concentrating solar power (CSP) systems, solar hot water (SHW), and solar process heat and space cooling. PV and SHW can be used in any location in the world. All new buildings should incorporate PVs and most new buildings should use SHW for heating domestic hot water. If new buildings do not incorporate solar technologies they should at least be made solar-ready so that solar technologies can be added easily in the future.

PV research and development encompasses: fundamental research in PV-related materials; the development of PV cells in several material systems; the characterization of PV cells, modules, and systems to improve performance and reliability; assisting industry with standardized tests and performance models for PV devices; and helping the PV industry accelerate manufacturing capacity and commercialization of various PV technologies.

04.18 / WIND-TURBINE-BLADE TESTING FACILITY.

04.19 / SOLAR PANELS ARE BEING DEVELOPED AT THE NATIONAL RESEARCH ENERGY LABORATORY.

04.20

Over the next 10 to 20 years the largest growth in research and development facilities will likely be in PV technology.

CSP technologies can be a major contributor to the future of the world's energy, particularly in areas that receive a large amount of solar energy. Many power plants use fossil fuels as a heat source to boil water. The steam from the boiling water spins a large turbine that drives a generator to produce electricity. However, a new generation of power plants with CSP systems use the sun as a heat source. The three main types of CSP systems are the linear concentrator, the dish/engine, and the power tower.

The linear-concentrator system collects the sun's energy using long rectangular, curved (U-shaped) mirrors. The mirrors are tilted toward the sun, focusing sunlight onto tubes (or receivers) that run the length of the mirrors. The reflected sunlight heats a fluid flowing through the tubes. The hot fluid is used to boil water in a conventional steam-turbine generator to produce electricity. There are two major types of linear concentrator systems: parabolic trough systems, where receiver tubes are positioned along the focal line of each parabolic mirror; and linear Fresnel reflector systems, where one receiver tube is positioned above several mirrors to allow the mirrors greater mobility in tracking the sun.

The dish/engine system uses a mirrored dish similar to a very large satellite dish. The dish-shaped surface directs and concentrates sunlight onto a thermal receiver, which absorbs and collects the heat and transfers it to the engine generator. The most common type of heat engine used today in dish/engine systems is the Stirling engine. This system uses the fluid heated by the receiver to move pistons and create mechanical power. The mechanical power is then used to run a generator or alternator to produce electricity.

04.20 / THE LINEAR-CONCENTRATOR SYSTEM USING PARABOLIC TROUGH COLLECTORS.

The power-tower system uses a large field of flat, sun-tracking mirrors known as heliostats to focus and concentrate sunlight onto a receiver on the top of a tower. A heat-transfer fluid

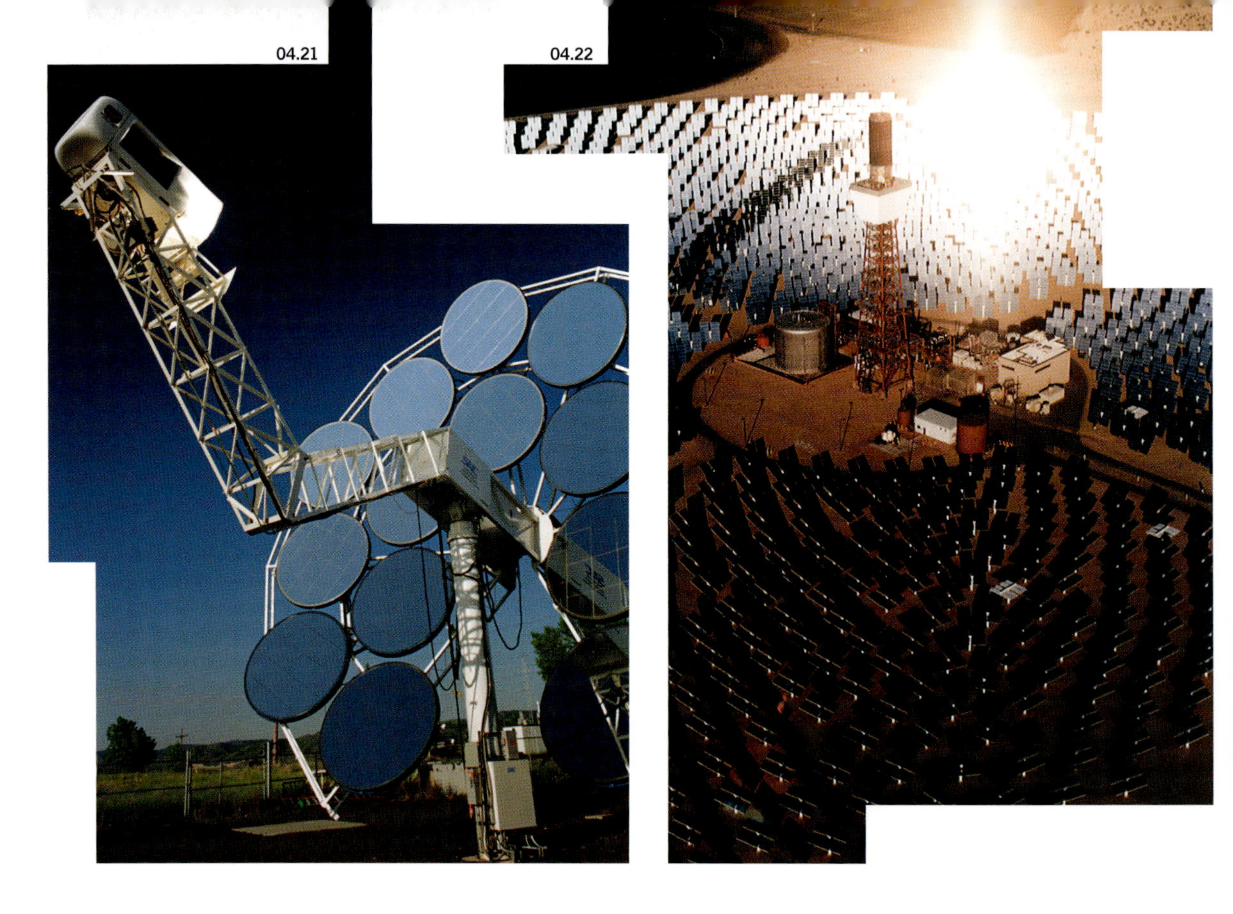

heated in the receiver is used to generate steam, which in turn is used in a conventional turbine generator to produce electricity. Some power towers use water/steam as the heat-transfer fluid. Other advanced designs are experimenting with molten nitrate salt because of its superior heat-transfer and energy-storage capabilities.

The linear-concentrator and power-tower systems have energy-storage capability, or thermal storage, that allows the systems to continue to dispatch electricity during cloudy weather or at night. The thermal storage is usually provided by using molten salt. The ability to incorporate large-scale energy storage is what makes CSP unique and important. Major CSP research areas are advanced materials development—including thermal storage fluids and optical materials—to support cost reduction and performance improvements.

Biomass /

Biomass is plant matter such as trees, grasses, agricultural crops, or other biological material. It can be used as a solid fuel, or converted into liquid or gas for the production of electric power, heat, chemicals, or fuels. By integrating a variety of biomass-conversion processes, all of these products can be made in one facility called a biorefinery. Biomass research and development at NREL is focused on: biomass characterization, especially cellulosic feedstocks; thermochemical and biochemical biomass conversion technologies; biobased-products development; and biomass process engineering and analysis. Many new biomass research facilities will be required in the next 20 years.

Hydropower /

Hydropower research aims to research, test, and develop innovative technologies capable of generating renewable and cost-effective electricity from water. These include marine

04.21 / THE DISH/ENGINE SYSTEM.

04.22 / THE SOLAR TWO PROJECT IMPROVES THE 10-MEGAWATT SOLAR ONE CENTRAL RECEIVER PLANT IN DAGGETT, CALIFORNIA. A FIELD OF MIRRORED HELIOSTATS FOCUSES SUNLIGHT ON A 300-FOOT TOWER, WHICH IS FILLED WITH MOLTEN-NITRATE SALT. THE SALT FLOWS LIKE WATER AND CAN BE HEATED TO 1050 °F.

04.23 / A GEOTHERMAL GREENHOUSE SHOWING EVAPORATIVE COOLING PADS—THE DARK WALL AT THE BACK OF THE GREENHOUSE.

04.24 / THIS 10 KW PV ARRAY WAS INSTALLED AT THE HYBRID POWER TEST BED/DER TEST FACILITY IN 2003. THE ARRAY, WHICH CONSISTS OF 96 BP SOLAR-POLYCRYSTALLINE-SILICON MODULES, IS USED TO TEST PV INVERTERS AND PACKAGED SYSTEMS.

and hydrokinetic technologies, which harness the energy from wave, tidal, current and ocean thermal resources, as well as technologies that improve the efficiency, flexibility, and environmental performance of conventional hydroelectric generation.

Geothermal /

Geothermal energy taps the heat that lies beneath the earth's surface in several ways. The geothermal reservoirs of hot water and steam are used to generate electricity and used for direct applications including aquaculture, crop drying, and district heating. Research on geothermal energy focuses on three areas: energy-systems research and testing (working to enhance conversion of geothermal energy into heat and electricity); drilling-technologies research (for both hardware and diagnostic tools); and geoscience and supporting-technologies research (exploration and resource management).

Electric infrastructure systems /

As the world's electric-power system ages, it is faced with increasingly difficult load and power demands. Physical, technical, and economic constraints have combined to place heavy burdens on an already taxed system. Research and development is required to strengthen the electric-power system through the integration of distributed energy resources and advanced power electronics. Major areas of research focus on: distributed energy testing and certification; interconnection standards and codes; interconnection and control technologies; energy-management, storage, and grid-support applications; and regulatory and institutional issues regarding how energy is distributed. ■

05. GLOBAL SUCCESS OF RESEARCH

As its name implies, genomics is the study of genes. Proteomics is the study of proteins produced by genes in cells, tissues, and organs. Both genomics and proteomics benefit greatly from bioinformatics, the use of advanced technology to develop and manage data. Advancements in these fields hold great promise. **In just a few years, the genomes of more than 20 of the world's deadliest disease-producing organisms, including those that cause malaria and tuberculosis, will be mapped. Researchers are now using this information to search for new diagnostics and treatments.**

Genomics /

Global collaboration has fostered the development of genomics—the first major area of research developed at a truly international level.

The Human Genome Project finished two years ahead of schedule largely because of supercomputers and international collaboration among educational institutions, government agencies, and private entities. The ambitious project involved six countries and enjoyed weighty financial backing. Leading participants were the $23-billion Wellcome Trust in the UK, and the National Institutes of Health, and Celera Genomics in the US.

Science /

Genetics is the study of inheritance or the way traits are passed down from one generation to another. Genomics, a newer term, is the study of all of one's genes, and the interactions of those genes with each other and a person's environment. The human genome is a person's complete set of DNA. Genes encode proteins, which direct cell activity and functions that affect health and influence traits, such as hair and eye color. To further understand how each gene specifically works, researchers are mapping the sites of action, fluctuations in levels, modifications, and interactions of proteins in cells.

DNA (deoxyribonucleic acid) is made up of four similar chemicals repeated millions or billions of times throughout a genome. The chemicals, called bases, are adenine, thymine, cytosine and guanine—abbreviated to A, T, C and G. The human genome has 3 billion pairs of bases. The particular order of As, Ts, Cs and Gs is extremely important. Each gene's code combines the four chemicals in various ways to spell out three-letter "words" that specify which amino acid is needed at every step in making a protein. The order underlies all of life's diversity, even dictating whether an organism is human or otherwise, such as yeast, rice or fruit fly, all of which have unique genomes and are themselves the focus of genome projects. Because all organisms are related through similarities in DNA sequences, insights gained from non-human genomes often lead to new knowledge about human biology.

Technology has revolutionized genomics. The genomes of each microbe may contain several million base pairs of DNA and thousands of genes. Whereas analyzing such an immense scale of data was previously daunting, the new DNA database and the computer infrastructure make it possible to run millions of samples a week.

For example, "Identification of the genes responsible for human Mendelian diseases (a disease due to mutations in a single gene)—once a Herculean task requiring large research teams, many years of hard work, and an uncertain outcome—can now be routinely accomplished in a few weeks by a single graduate student with access to DNA samples and associated phenotypes, an Internet connection to the public genome databases, a thermal cycler, and a DNA-sequencing machine," reports the National Human Genome Research Institute (NHGRI), a component of the National Institutes of Health.[1] In fact, supercomputers now create so much data that instead of spending a great deal of time generating data, researchers must now carefully choose which data to pursue among the abundant supply.

Opportunities /

Genomics raises the prospect of personalized medicine. Researchers expect to be able to identify the specific markers that distinguish an individual's disease and create a treatment to correct it. By identifying a person's specific gene variations, the adverse side effects of non-personalized medications can be reduced and treatments can be improved. An example of this kind of targeted medicine is the anti-cancer drug Gleevec, which has shown to be effective with patients who have a particular genetic mutation.

Since the completion of the Human Genome Project in 2003, the National Human Genome Research Institute has launched many exciting research initiatives aimed at improving human health. These projects include the International HapMap Project, the ENCyclopedia of DNA Elements (ENCODE), and a chemical genomics initiative. The NHGRI hopes to cut the cost of whole-genome sequencing to under $1,000, which will allow individual genomes to be sequenced as part of routine medical care to help medical professionals predict disease, personalize treatment, and preempt illnesses.

Genomics has applications beyond the study of genes. Recent increases in chronic diseases such as diabetes, childhood asthma, obesity, and autism are linked to environmental toxins or poor diets and lack of exercise. Researchers are using new tools of genomics and proteomics and high-tech sensors to measure environmental toxins, dietary intake, physical activity, and metabolism rates to determine an individual's biological response to such influences.

05.1 / DNA, WITH ITS DOUBLE HELIX SHAPE, IS THE WORLD'S MOST RECOGNIZABLE MOLECULE.

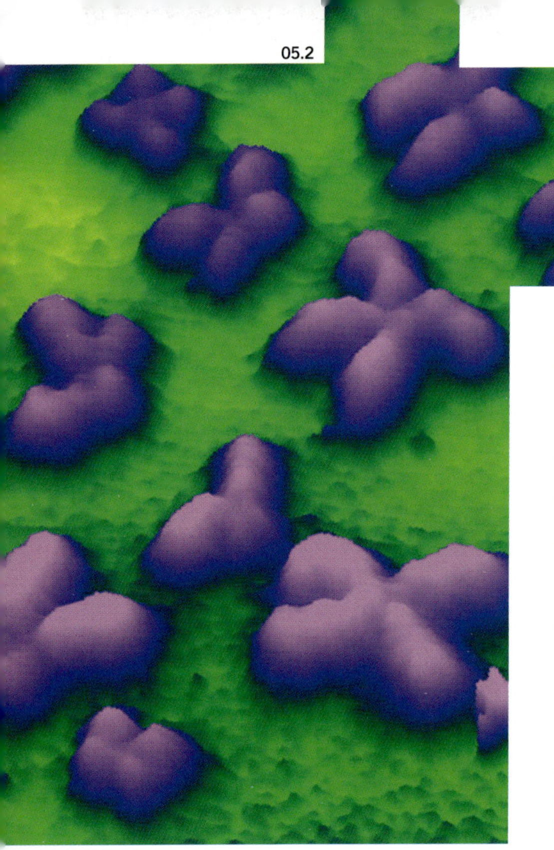

05.2

Other opportunities for genomics research include DNA forensics to help identify:

- Crime suspects
- Crime and catastrophe victims
- Organ donors for transplant recipients
- Paternity and other family relationships
- Protected species (could be used for prosecuting poachers)
- Pedigree for seed or livestock breeds.

Global collaboration in mapping the human genome was a great success, but these types of partnerships raise challenges in areas such as intellectual property rights. The Genome Project produced vast amounts of rich data available to anyone through the Internet. As we navigate this new frontier, however, international laws and practices must be developed and supported so that the information is available to anyone who wants to use it.

Personalized Medicine /

Advances brought about by genomics have allowed researchers to decipher the genetic instructions encoded in the estimated 3 billion base pairs of nucleotide bases that make up human DNA. Analysis of this information is revolutionizing our understanding of how genes control the functions of the human body. It is also helping to explain the mysteries of embryonic development, empowering us to discover the indicators and causes of diseases and their treatment.

In 2009 Dr. Leroy Hood, a world-leading molecular-biotechnology and genomics scientist and co-founder of the Institute of Systems Biology in Seattle, Washington, gave an insightful presentation to 50 architects at Perkins+Will. Our architectural firm designs research facilities around the world. Dr. Hood's message was very clear, "**today medicine is reactive, tomorrow (over the next 10 years) medicine will be proactive.**" He explained his P4 solution, which focuses on predictive, personalized, preventative, and participatory medicine.

The cost today for a personalized test based on DNA is about $10,000. Soon technology and competition should bring the cost of a test down to about $1,000. The test should be able to predict most potential diseases. This process will be proactive to help prevent many problems occurring. The new healthcare will provide patients with the opportunity to address their specific needs much sooner in life. There will need to be a steep educational process for doctors, pharmaceutical companies, and the entire healthcare industry so that data can be interpreted and explained clearly to patients. Diagnosis will become very important in the near future. Dr. Hood also pointed out that **throughout history "each new idea needed a new organizational structure."** We should soon be at that point as research and technology continue to evolve at a rapid pace.

Other key points shared by Dr. Hood included a discussion on **systems biology and the need to integrate scientific research.** He said the interdisciplinary research in biology, technology, and computational processing should be addressed in a holistic manner. For example, systems information is the combination of digital information, which is generated from DNA modeling, and environmental research, which focuses on living organisms and diseases. Leading medical schools around the world are adopting systems research to support the study of medicine.

05.3

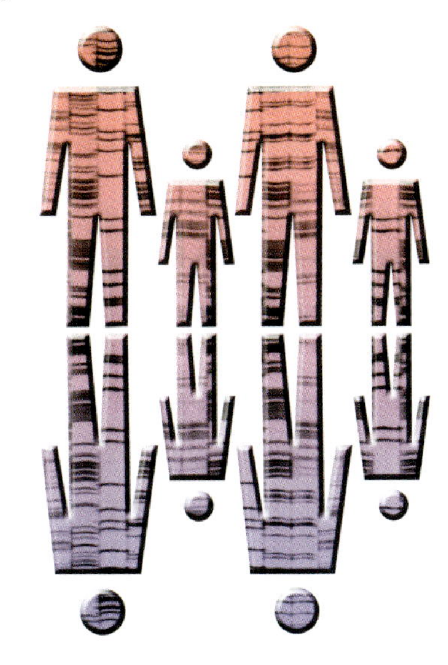

Dr. Hood's group is also doing important work on disease diagnostics. There are specific blood-protein fingerprints that, for instance, detect the presence of a cancer eight to ten weeks earlier than is possible by way of clinical diagnosis alone, thereby allowing for faster and more effective treatment. For example, 15 percent of prostate-cancer cases are malignant and need to be addressed quickly and aggressively, while the other 85 percent do not need to be treated. This results in an estimated saving of $20 billion per year for just one disease. **The technology to support proactive medical treatment is expected to grow exponentially over the next 10 years, and blood diagnostics will be a key area of focus.**

Biobanks /

As an integral component of the data-collection process, biobanks represent the leading edge of this new frontier of biomolecular research. These highly specialized facilities are designed for the collection, storage, processing, and distribution of human biological materials and their corresponding analytical data.

Biobanks allow researchers to integrate collections of bio-specimens (blood, DNA, tissue, biopsy specimens, etc.) with corresponding patient data such as genetic profiles, medical histories, and lifestyle information. By combining and comparing biological samples with genetic information and medical histories, researchers are able to investigate the fundamental mechanisms of diseases in rich new ways. New insights into molecular and genetic processes will lead to better techniques for predicting who is susceptible to particular illnesses by hunting down the genetic processes that cause them, as well as discovering better and more innovative ways to target and treat many diseases.

05.2 / HUMAN CHROMOSOMES IN METAPHASE.

05.3 / A GRAPHIC REPRESENTATION OF DNA "FINGERPRINTS".

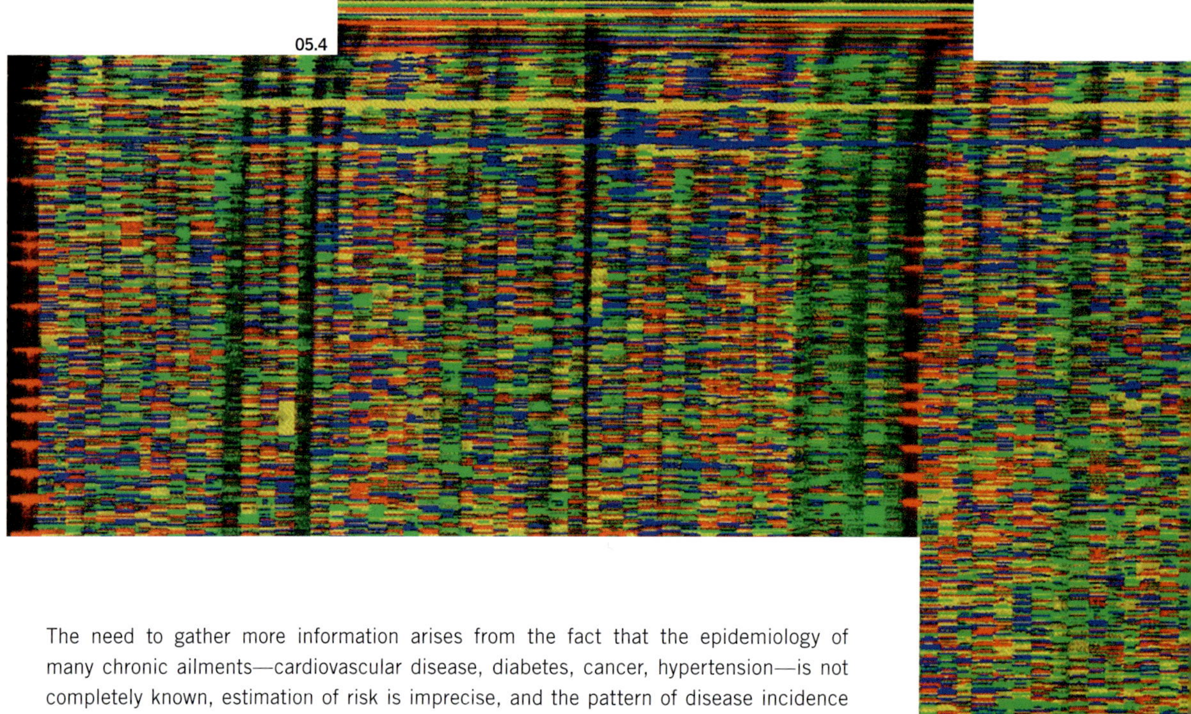

05.4

The need to gather more information arises from the fact that the epidemiology of many chronic ailments—cardiovascular disease, diabetes, cancer, hypertension—is not completely known, estimation of risk is imprecise, and the pattern of disease incidence is changing—and these changes need to be documented. **Changing patterns of lifestyle, such as smoking, nutrition and exercise, as well as advances in medical care, may all influence future morbidity and mortality rates.** As changes in early detection and treatment advance, prospective epidemiology is needed to document the value and impact of these changes in an organized fashion.

Currently, there are a number of biorepositories around the world maintained by various institutions, however, most of these are relatively small collections containing a few hundred to a few thousand samples, and are related to specific diseases or to narrow focus programs. While these are relevant and important, it is the large-scale, population-based biobanks that facilitate study of not only a single disease, but also the effect of genes in a wide range of complex systems and health-related issues. The key to a powerful comprehensive database is high-quality samples from as wide range of the population as possible.

For the Kingdom of Saudi Arabia, a relatively young country (established in 1932) with a young population characterized by large family size, it has become an objective to quantify disease incidences in various Saudi populations and subpopulations. The Saudi research community recognizes that biobanks are a good tool to developing an understanding of natural histories and genome-environment interactions in order to translate the risk factors for endemic diseases.

A variety of study designs have been identified that can be used to investigate different aspects of the relationships between different exposures and the risk of disease. These have included family-based studies of genetic factors, retrospective case-control studies of particular conditions, and prospective observational studies. However, for the sake of comprehensive and reliable quantification of the combined affects of lifestyle, environment, genotype, and other exposures of a variety of outcomes, Saudi Arabia has recognized a strategic need to establish some large blood-based prospective epidemiological studies that can only be facilitated by the use of biobanks.

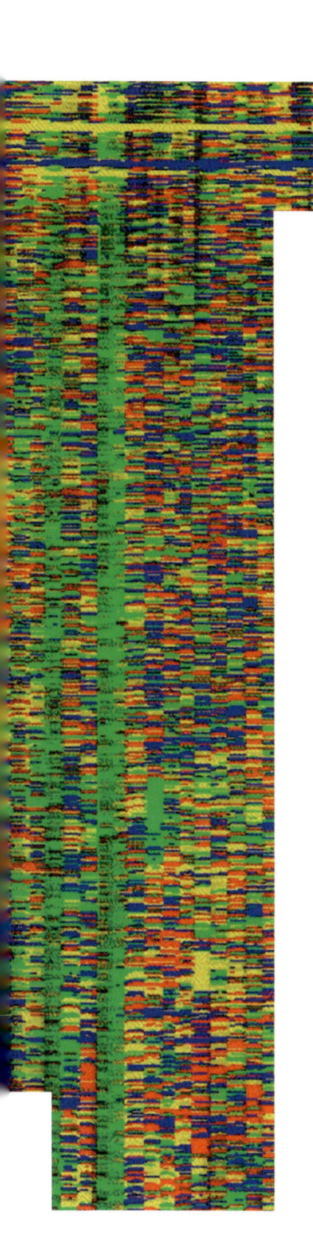

These studies will be concerned with two groups:

1. A prospective family-based group, with an expected sample of 100,000 people of all ages comprising about 20,000 to 25,000 families, to examine these interactions in a range of common late-onset diseases.
2. A disease-specific group, where samples of people affected by certain diseases such as cancer, diabetes, hypertension, etc. (around 100,000 cases) will be allocated from specialized clinics in Saudi Arabia.

Researchers are aware that having so many samples in one database unquestionably boosts the power of genetic research, but it also raises some ethical questions: How should biobanks gain consent from donors? Can they ensure that each person's details remain private and confidential? Who has the rights to the benefits, or sharing, of intellectual property—especially as access to such a database will be attractive to researchers and agencies from a range of organizations? With these concerns in mind, it has being considered that participants should have the right to withdraw from a study at any time, and may allow the project to continue making use of the DNA information already provided, or may have all their data destroyed. It is clearly understood that the effectiveness, efficiency, and long term potential of this establishment depends significantly on public perception, and the people's willingness to contribute to its collection. The willingness to contribute is mainly driven by altruism, and depends on the public being well informed and having trust in the experts and institutions.

The goal of the biobanks is to increase the quality of patient care and to accelerate the impact of research on that care. They will implement the highest standards of biological banking to provide outstanding clinical, medical, demographic, and analytical data, with the objective to:

- Encourage the donation of specimens
- Educate the public about the advancements that could come from these studies
- Enable the biorepository community to collaborate on implementing the best practices and most promising approaches
- Enable the quantification of disease incidences in various populations and sub-populations
- Enable an understanding of natural histories and risk factors for these diseases, including genome-environment interactions
- Accelerate the discovery and development of new diagnostics and therapeutics.

I have been involved in the design of four biobanks in Saudi Arabia. The biobank holds 500,000 samples at –80 °F in a 20-foot-high by 45-foot-long freezer for several decades. This is one of the most visually impressive pieces of scientific equipment I have seen. I also had the opportunity to walk into the freezer for two minutes, at which point I could start to feel my body freezing.

05.4 / AUTOMATED DNA SEQUENCING OUTPUT.

History /

The Human Genome Project began in 1990, and in February 2001 *Nature* published the International Human Genome Sequencing Consortium's first draft of the human genome, with the sequence of the entire genome's 3 billion base pairs 90 percent complete.

In April 2003, on the 50th anniversary of the discovery of the double-helical structure of DNA, the Humane Genome Project presented a high-quality, comprehensive sequence of the human genome that was 99.9 percent complete.

In collaboration with the Association of Schools of Public Health, the CDC established the first Centers for Genomics and Public Health in 2001. Located in schools of public health at the universities of Michigan, North Carolina, and Washington, these hubs worked with existing university programs and linked state and local health departments. In 2003, the CDC launched partnerships with the state health departments of Michigan, Minnesota, Oregon, and Utah to support genomic research into chronic diseases.

There are three immediate priorities for public health research in genomics:

1. Understanding how genomic factors influence the health of populations.
2. Examining the value of genetic tests for screening and prevention.
3. Assessing the value of family medical history.

In 2005, a NHGRI-led consortium of scientists from six countries published the haplotype map or HapMap. This map of the human genome may prove even more powerful than the human genome sequence. While, the human genome sequence identified 99.9 percent of the DNA code that we all share, the HapMap revealed in detail the 0.1 percent that represents genome variation. This powerful tool is uncovering differences that predispose some people to diseases, ranging from heart disease, diabetes, Alzheimer's, and cancer. Like the Human Genome Project, all of the HapMap data is publicly available on the Internet.

The unrestricted access to the genome mapping on the Internet has resulted in over a million hits each week.

Funding /

Funding is critical for research. The Wellcome Trust, the world's second-largest privately supported trust (next to the Bill and Melinda Gates Foundation), supports genomics research to the tune of nearly $1 billion annually. The trust was established in 1936 upon the death of Henry Wellcome, an American pharmacist who co-founded a new form of compressed pill. Wellcome moved to England, where he set up several research laboratories and employed leading scientists. In 1924, he consolidated his commercial and non-commercial activities under one corporate umbrella: The Wellcome Foundation Limited. In 1932, he was knighted and made an Honorary Fellow of the Royal College of Surgeons of England. Today, the Wellcome Genomics Campus is one of the world's most significant research facilities.

The budget for the National Human Genome Research Institute is less than 2 percent of NIH funding ($483 million for 2007). In partnering with other institutions globally, the NHGRI benefits from money invested in genomic research internationally.

Funding for research must be thoughtfully applied. There is a tendency to support the next new thing, often at the expense of further developing existing data archives, for example. The result is a tremendous loss of research potential. Research funding must be coordinated to be as effective and as efficient as possible.

Proteomics /

Genomics—used to sequence of the genomes of humans and other organisms—has led to proteomics. The study of proteins requires much more data, which has spawned a new type of research: bioinformatics.

Science /

Whereas genomics is the study of the human genome itself, proteomics is the study of proteins expressed (produced) by those genes. Proteomics studies how groups of proteins in cells, tissues, and organs respond, interact, and change. If sequencing the human genome was a massive undertaking, the challenge of creating a map of the human proteome is astronomical. Even the lowly fruit-fly gene has been found to code for 38,000 different proteins. "So while we now know that humans have approximately 30,000 genes, our total number of proteins is still unknown—although estimates put the total at 100,000 or more," reported Science.org.[2]

The Science.org article went on to point out that, "while the genome remains relatively static in any given cell in an organism, the expression of the proteome changes from cell to cell and from moment to moment. Age, gender, health, and recent consumption of food or drugs all affect the proteome—the same cell, if examined at different

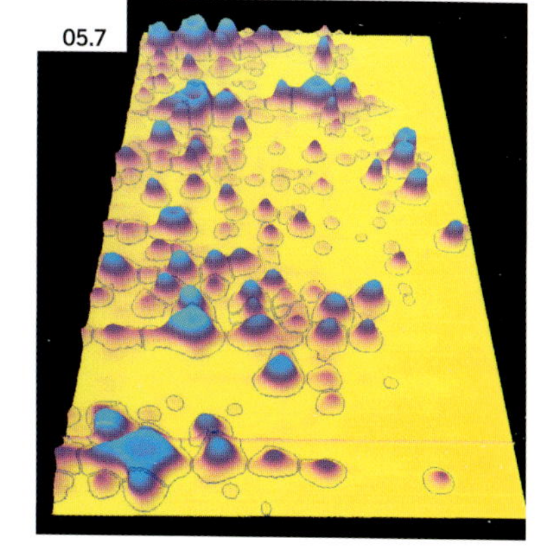

05.5 / THE UK BIOBANK FREEZER STORES 500,000 SAMPLES.

05.6 / A SELECTION OF SAMPLES IN THE UK BIOBANK.

05.7 / PROTEOMICS—THE SEPARATED PROTEINS FORM PART OF THE BRAIN MITOCHONDRIAL PROTEOME.

times or under different conditions, can be expressing a different complement of proteins." [3]

Proteomics is particularly important because most diseases are manifested at the level of protein activity. Proteomics seeks to correlate directly the involvement of specific proteins, protein complexes, and their modification status in a given disease state. Such knowledge will speed up the identification of new drug targets that can be used to diagnose and treat diseases. In the next five to 10 years, genomics is expected to produce many drug discoveries, followed by a large wave of discoveries made possible through proteomics.

Opportunities /

Mapping of the human genome has paved the way for the study of proteins through proteomics. Genes are not agents of biochemical activity and they do not directly interact with each other—they are essentially a passive information store. In contrast, proteins do interact with and modify each other and many other biochemicals, including DNA. Also, genes are biochemicals (DNA) whose information is stored in a single dimension—the order of the bases that make up their sequence. Proteins, on the other hand, catalyze biochemical functions and the transmission of cellular signals.

Advancements in proteomics will require significant developments in technology, instruments, standardized tools, protocols, and software to detect, catalog, and analyze the huge amounts of data. It will also require international collaboration—a large-scale, coordinated effort among multiple laboratories linked with powerful informatics capabilities. To that end, the National Cancer Institute in 2005 launched the Clinical Proteomic Technologies Initiative for Cancer. The five-year program is:

- Networking multiple research laboratories to permit large-scale, real-time exchange and application of protein measurement technologies, biological resources, and data dissemination;
- Refining and standardizing technologies, reagents, methods, and analysis platforms to ensure reliable and reproducible separation, capture, identification, quantification, and validation of protein measurements from complex biological mixtures;
- Evaluating approaches to separate and recognize proteins involved in cancer development.

On an international level, the Human Proteome Organization (HUPO) is a consortium of proteomics research associations, government researchers, academic institutions, and industry partners. HUPO promotes proteomics research and facilitates scientific collaborations.

05.9

05.10

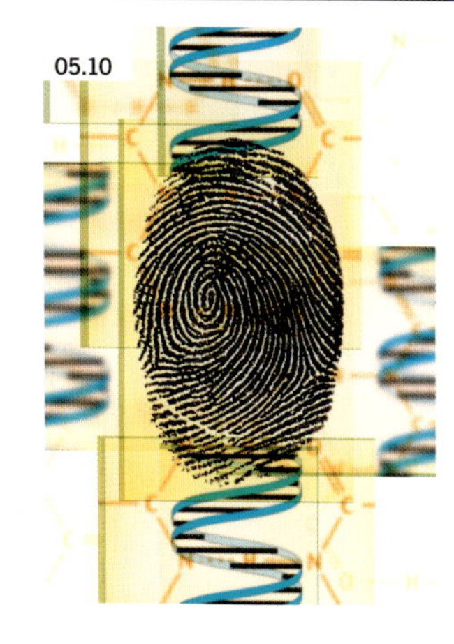

Bioinformatics /

Bioinformatics is facilitating biological discoveries and creating a global database from which unifying principles in biology can be studied. Through its applications in research fields such as genomics and proteomics, bioinformatics is used in drug discovery, molecular medicine, microbial genome applications, agriculture, comparative studies, and other areas of life science. The field of bioinformatics has evolved such that the most pressing task now involves the analysis and interpretation of various types of data. The actual process of analyzing and interpreting data is referred to as computational biology.

Bioinformatics merges biology, computer science, and information technology to form a single discipline. Bioinformatics is integral to genomic and proteomic research.

Bioinformatics is trying to address three key points: first, the need to capture and integrate the data; second the needs for an efficient means of accessing, reviewing, and analyzing the data; and third, the data and conclusions should be shared. This field will grow rapidly in coming years because computer technology will make it possible to crunch even larger amounts of data faster and more effectively.

The new centers for proteomics will share ideas and develop faster ways to bring discoveries to market. They will also support advancements in the following areas:

- Protein profiling, which quantifies a large number of different proteins in order to reveal molecular pathways;
- Post-translation modifications, which examine how modifying a protein's structure alters its function;
- Protein interactions, which look at how proteins interact with themselves and various cellular factors.

Computers and automated equipment are critical to science, and will become even more critical as bioinformatics develops and as research transitions from chemicals and fume hoods to computer modeling and data crunching. The combined effect of this huge influx of data is collaboration and intellectual openness.

Genomics and proteomics have the potential to help us understand future health risks and to address those problems proactively to improve the quality of life, manage healthcare costs more effectively, and extend life beyond what is possible today.

05.8 / A HIGH-THROUGHPUT CELL-CULTURE ROBOT.

05.9 / BIOINFORMATICS—GENE-EXPRESSION ANALYSIS.

05.10 / GRAPHIC OF DNA "FINGERPRINTING".

06. POLITICAL SCIENCES

Introduction /

Embryonic stem cell research is flourishing on an international scale. In the US it remains controversial, although a solid majority of Americans now support it.[1] Opposition can slow the process, but it will not stop the research in private industry or in other countries. Europe has taken the lead in stem-cell research, although Asia isn't far behind, and as a result it will likely be the first to benefit medically and financially.

In March 2009 President Obama announced: "We will lift the ban on federal funding for promising embryonic stem-cell research. We will also vigorously support scientists who pursue this research. And we will aim for America to lead the world in the discoveries it one day may yield. At this moment, the full promise of stem-cell research remains unknown, and it should not be overstated. But scientists believe these tiny cells may have the potential to help us understand, and possibly cure, some of our most devastating diseases and conditions: to regenerate a severed spinal cord and lift someone from a wheelchair; to spur insulin production and spare a child from a lifetime of needles; to treat Parkinson's, cancer, heart disease, and others that affect millions of Americans and people who love them."

Stem Cells /

Stem Cell Science /

Stem cells—the parent cells for all tissues and organs of the body—exist primarily to maintain and repair cells. Found in blood, bone marrow, muscles, and organs, such as the brain, liver, and skin, stem cells have unique characteristics. First, they can continually renew themselves through cell division. Second, they can differentiate, or transform into specialized cells with specific functions, such as a heart cell or a nerve cell.

Scientists work with both animal and human stem cells as well with both embryonic and adult stem cells. **Embryonic stem cells can truly become any cell type of the body,** making them more versatile than adult stem cells.

Because of their theoretical potential to replace, repair, and regenerate cells and whole organs, embryonic and even adult stem cells stir great excitement among researchers and others worldwide. Some key researchers are relocating to areas where stem-cell research is well funded. As this brainpower shifts, so shifts the economic outlook of nations.

06.2

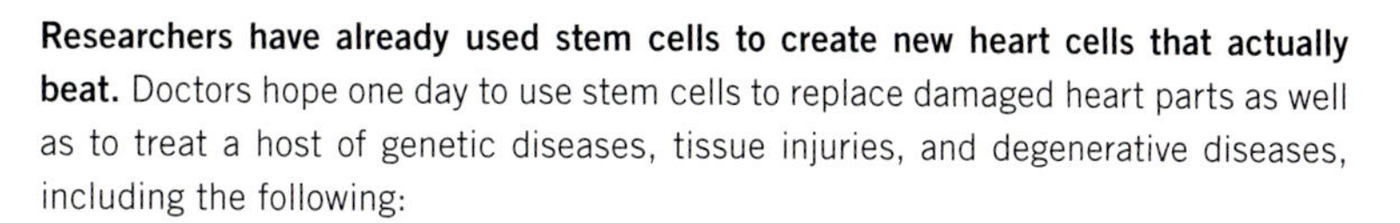

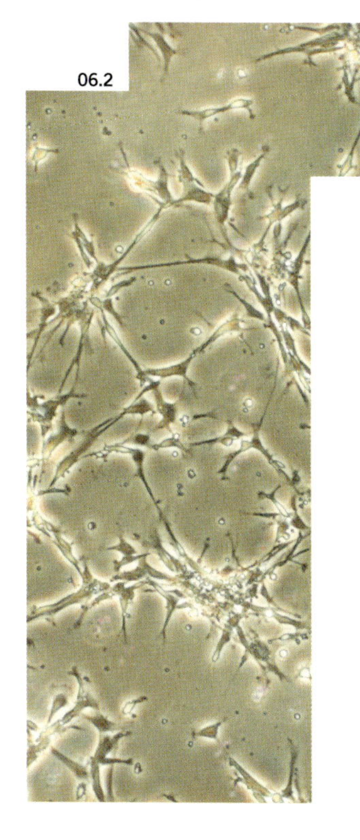

Researchers have already used stem cells to create new heart cells that actually beat. Doctors hope one day to use stem cells to replace damaged heart parts as well as to treat a host of genetic diseases, tissue injuries, and degenerative diseases, including the following:

- Alzheimer's
- Cancer
- Birth defects
- Burns
- Craniofacial abnormalities
- Diabetes
- Fanconi anemia
- Heart disease
- Hemophilia
- Infertility
- Liver disease
- Macular degeneration
- Muscular dystrophy
- Parkinson's disease
- Poor circulation
- Pulmonary disease
- Sickle cell disease
- Spinal cord injury
- Stroke

In the US alone, nearly 100 million people—almost one-third of the population—suffer from one of these devastating conditions for which treatments are sought.[2]

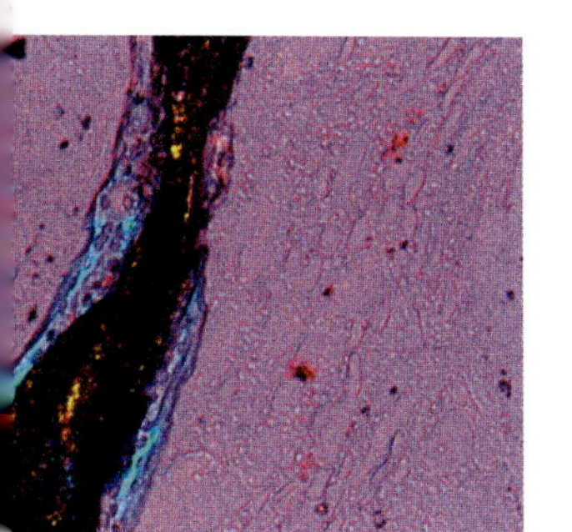

Embryonic Stem Cells /

Research involving stem cells derived from human embryos has prompted a great deal of ethical debate. Human embryonic stem cells are generated from fertilized, frozen eggs at in vitro fertilization clinics. Informed donors donate these embryos to research because they decide against using them to become pregnant.

Embryonic stem cells are derived from four- to five-day-old embryos known as blastocysts. Each blastocyst consists of 50 to 150 cells and includes three structures: an outer layer of cells, a fluid-filled cavity, and a group of about 30 pluripotent cells at one end of the cavity. The cells in the latter group, called the inner cell mass, form all of the body's cells.

In his book *Lucky Man*, actor and stem cell research advocate Michael J. Fox, who suffers from Parkinson's disease, makes the following case for why embryonic stem-cell research should be aggressively pursued: "Embryonic stem cells are taken from ... embryos left over from in vitro fertilization and discarded by fertility clinics. Thousands of these unwanted cell clusters, smaller than the head of a pin, are frozen and then, after a time, routinely destroyed every year." Because these cells have not yet dedicated themselves to any one physiological function, they are pluripotent—they have the potential to become any type of human cell. "Introduced into the substantia nigra of a Parkinson's patient, for example, they could evolve into dopamine-producing cells," Fox states.[3]

The use of stem cells also fuels controversy because it is linked with cloning. Although most stem-cell research does not involve cloning, the two fields are often

06.1 / NEW BONE GROWTH WITHIN A BIOCERAMIC BONE SUBSTITUTE.

06.2 / HUMAN NEURAL STEM CELLS CAN BE MADE TO DEVELOP INTO CELLS FOUND IN THE CENTRAL NERVOUS SYSTEM.

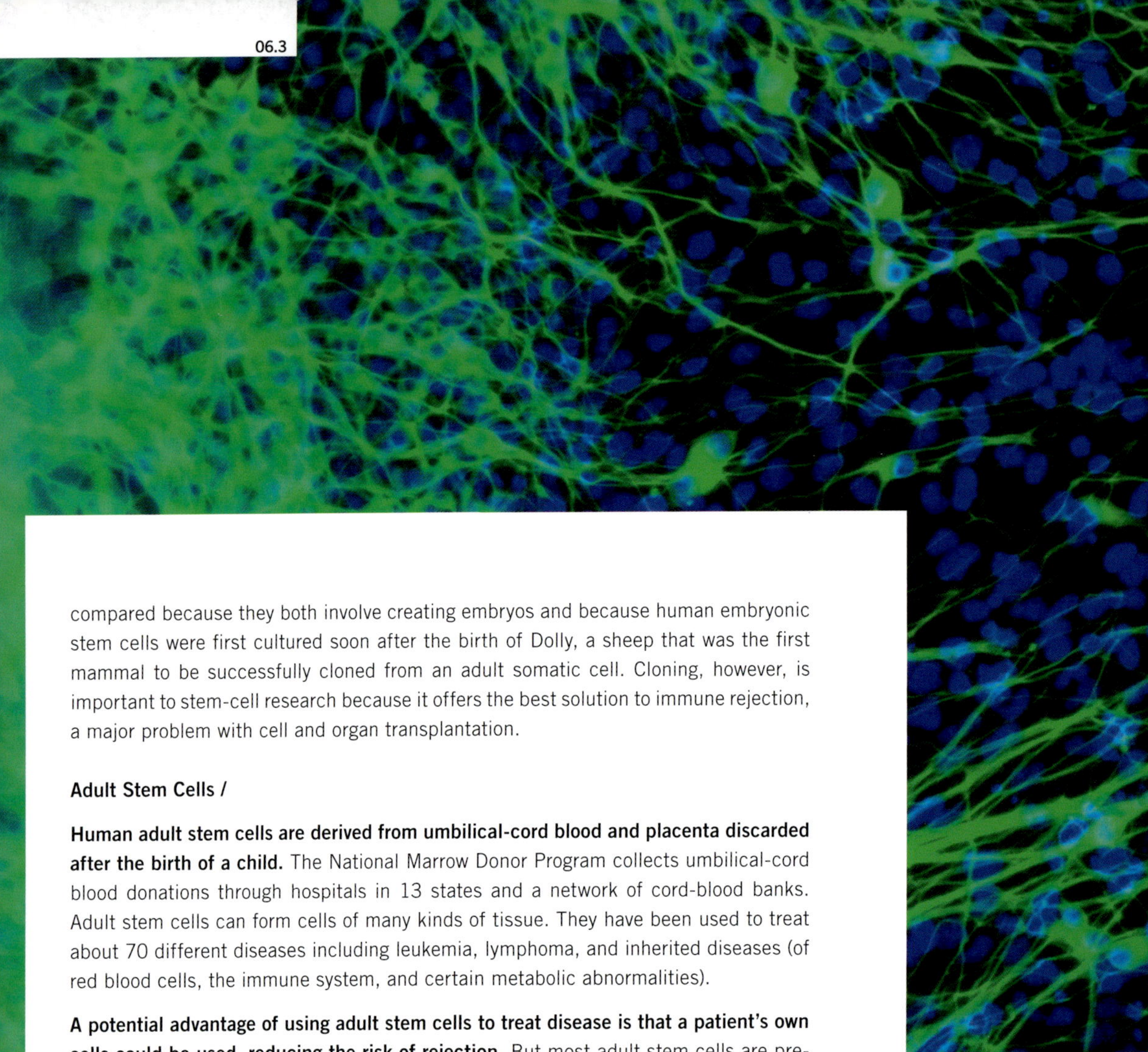

compared because they both involve creating embryos and because human embryonic stem cells were first cultured soon after the birth of Dolly, a sheep that was the first mammal to be successfully cloned from an adult somatic cell. Cloning, however, is important to stem-cell research because it offers the best solution to immune rejection, a major problem with cell and organ transplantation.

Adult Stem Cells /

Human adult stem cells are derived from umbilical-cord blood and placenta discarded after the birth of a child. The National Marrow Donor Program collects umbilical-cord blood donations through hospitals in 13 states and a network of cord-blood banks. Adult stem cells can form cells of many kinds of tissue. They have been used to treat about 70 different diseases including leukemia, lymphoma, and inherited diseases (of red blood cells, the immune system, and certain metabolic abnormalities).

A potential advantage of using adult stem cells to treat disease is that a patient's own cells could be used, reducing the risk of rejection. But most adult stem cells are pre-specialized, that is, blood stem cells make only blood, and brain stem cells make only brain cells. However, as discussed in the following section, major discoveries in late 2007 showed that adult stem cells can be reprogrammed to behave just like embryonic stem cells that can develop into any type of cell. And prior to those findings, University of Minnesota researchers had shown that some adult stem cells in bone marrow can, under the right laboratory conditions, develop into other kinds of cells, and may therefore be much more potent.[4] Adult stem cells are scarcer in the body and harder to grow in the lab than embryonic stem cells, and they don't seem to be as versatile.

Stem Cell Breakthroughs /

In the fall of 2007 a major stem-cell breakthrough occurred when two groups of scientists independently demonstrated that **they were able to reprogram human adult stem cells to act like embryonic stem cells, thus changing the direction of stem-cell research and signaling that future research will likely not be dependent upon embryos and eggs.** "This is like an earthquake for both science and politics of stem cell research," Jesse Reynolds, policy analyst for the Center for Genetics and Society, told ScienceNOW.[5]

06.4

One group of scientists, led by Dr. Shinya Yamanaka of Kyoto University, tricked a cheek cell from a middle-age woman into acting like an embryonic stem cell. The other group, led by James Thomson, the molecular biologist at the University of Wisconsin who in 1998 discovered the first human embryonic stem cells, reprogrammed human cells from fetal skin and from the foreskin of a newborn boy. In both cases, the created cells were truly pluripotent, meaning they had the potential to grow into any tissue in the body, reported ScienceNOW.[6]

Yamanaka told *Time* magazine, **"We can now for sure begin to generate patient-specific cells.** And we should be able to use them in cell-transplant therapy."[7] According to the Time report, "Other giants of the field seem to agree." Ian Wilmut, the scientist who cloned Dolly the sheep, has abandoned "the egg-hollowing, gene-replacing technique that made him a pioneer, noting that 'changing cells from a patient directly into stem cells has got so much more potential.' No embryos, no eggs, no hand-wringing over where the cells came from and whether it was ethical to make them in the first place."

The pluripotent stem cells created by both team of scientists can be transplanted into the donor with little risk of rejection, just like embryonic stem cells. "I think this is the future of stem-cell research," Dr. John Gearhart, the biologist from John Hopkins University who first isolated human fetal embryonic stem cells, told *Time*.

Until more studies are conducted, however, scientists are not yet ready to completely abandon stem cells from embryos. But given the recent breakthroughs, that day is not far away, they suspect. "Embryonic stem-cell technology is already looking rather last-century, along with therapeutic cloning," wrote Dr. Patrick Dixon, often described as Europe's leading futurist, even prior to the November 2007 stem-cell breakthrough.[8]

"History will show that by 2020 we were already able to produce a wide range of tissues using adult stem cells, with spectacular progress in tissue building and repair. In some cases these stem cells will be actually incorporated into the new repairs as differentiated cells, in other cases, they will be temporary assistants in local repair processes. We will also see some exciting new pharmaceutical products in the pipeline, which promise to do some of the same tricks without having to remove a single stem cell from the body. These drugs may, for example, activate bone marrow cells and encourage them to migrate to parts of the body where repairs are needed," stated Dixon on globalchange.com.

06.3 / NEURONS (GREEN) DERIVED FROM EMBRYONIC STEM CELLS IN CULTURE.

06.4 / THE GREEN STAINING HIGHLIGHTS THE TYPICAL APPEARANCE OF DIFFERENTIATED NERVE CELLS. CELLS SUCH AS THESE COULD FORM THE BASIS OF FUTURE TREATMENTS FOR DEGENERATIVE BRAIN DISEASES SUCH AS PARKINSON'S AND ALZHEIMER'S.

Key Developments in Stem-Cell Research /

Stem-cell research is developing at such a rapid pace that over 2,000 new research papers are published in reputable scientific journals each year. Below, sourced in part from NPR.org[9], is a timeline of key developments in this promising new science. All the breakthroughs within the past 10 years illustrate how quickly stem-cell research is evolving.

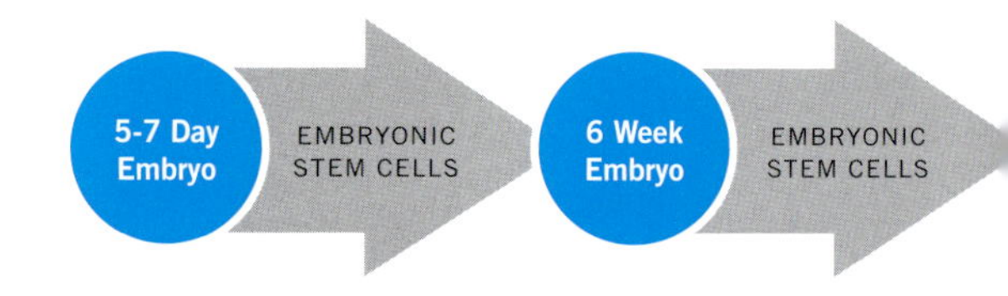

06.5 / THE AGE OF VARIOUS TYPES OF STEM CELLS

1981 EMBRYONIC STEM CELLS ARE FIRST ISOLATED IN MICE.

1995 UNIVERSITY OF WISCONSIN RESEARCHERS ISOLATE THE FIRST EMBRYONIC STEM CELLS IN PRIMATES, PROVING THAT IT'S POSSIBLE TO DERIVE EMBRYONIC STEM CELLS FROM HUMANS.

1996 THE FIRST MAMMAL IS CLONED, A SHEEP BORN IN SCOTLAND CALLED DOLLY.

1998 SCIENTIST RICHARD SEED SAYS HE WILL OPEN A HUMAN-CLONING CLINIC. THE CLINIC NEVER OPENS, BUT IT STIRS DEBATE ON THE ETHICS OF HUMAN CLONING.

RESEARCHERS AT THE UNIVERSITY OF WISCONSIN AND JOHNS HOPKINS UNIVERSITY REPORT ISOLATING HUMAN EMBRYONIC STEM CELLS. THE CELLS HAVE THE POTENTIAL TO BECOME ANY TYPE OF CELL IN THE BODY AND MIGHT ONE DAY BE USED TO REPLACE DAMAGED OR CANCEROUS CELLS.

2000 THE US NATIONAL INSTITUTES OF HEALTH ISSUE GUIDELINES THAT ALLOW FEDERAL FUNDING OF EMBRYONIC STEM-CELL RESEARCH. FORMER PRESIDENT BILL CLINTON SUPPORTS THE GUIDELINE.

2001 A MONTH AFTER TAKING OFFICE, PRESIDENT GEORGE W. BUSH PUTS A HOLD ON FEDERAL FUNDS FOR STEM-CELL RESEARCH.

SENATOR BILL FRIST (R-TN), A RESPECTED SURGEON, AND SENATOR ORRIN HATCH (R-UT) CALL FOR LIMITED FEDERAL FUNDING FOR STEM-CELL RESEARCH.

PRESIDENT BUSH SAYS HE WILL LIMIT FUNDING TO A FEW DOZEN LINES OF EXISTING EMBRYONIC STEM CELLS. MANY OF THOSE LINES LATER PROVE TO BE CONTAMINATED, AND SOME CONTAIN GENETIC MUTATIONS, MAKING THEM UNSUITABLE FOR RESEARCH.

2004 NEW JERSEY LEGISLATORS AUTHORIZE $9.5 MILLION FOR A STEM-CELL INSTITUTE, MAKING NEW JERSEY THE FIRST STATE TO FUND RESEARCH ON STEM CELLS, INCLUDING THOSE DERIVED FROM HUMAN EMBRYOS.

CALIFORNIANS APPROVE PROPOSITION 71 AUTHORIZING THE STATE TO SPEND $3 BILLION ON EMBRYONIC STEM-CELL RESEARCH OVER 10 YEARS. IT PUTS CALIFORNIA AHEAD OF THE FEDERAL GOVERNMENT AND MANY OTHER NATIONS IN PROMOTING THE RESEARCH.

2005 MORE THAN 170 BILLS ON STEM-CELL RESEARCH (BOTH FOR AND AGAINST) WERE CONSIDERED BY STATES ACROSS THE USA.

2005 RESEARCHERS AT THE UNIVERSITY OF WISCONSIN-MADISON TRANSFORM EMBRYONIC STEM CELLS INTO SPINAL MOTOR NEURONS.

THE HOUSE PASSES A BILL THAT WOULD EASE PRESIDENT BUSH'S RESTRICTIONS ON FEDERAL FUNDING FOR STEM-CELL RESEARCH.

CONNECTICUT APPROVES $100 MILLION IN FUNDING FOR ADULT AND EMBRYONIC STEM-CELL RESEARCH OVER THE NEXT 10 YEARS.

BYPASSING THE ILLINOIS STATE LEGISLATURE, DEMOCRATIC GOVERNOR ROD BLAGOJEVICH CREATES A STEM-CELL-RESEARCH INSTITUTE BY EXECUTIVE ORDER. THE INSTITUTE IS TO BE FUNDED THROUGH A LINE ITEM IN THE STATE BUDGET THAT GIVES THE PUBLIC HEALTH DEPARTMENT $10 MILLION TO FUND RESEARCH.

IN DEFIANCE OF PRESIDENT BUSH, SENATE MAJORITY LEADER BILL FRIST (R-TN) ANNOUNCES HIS SUPPORT OF LEGISLATION TO EASE FEDERAL FUNDING RESTRICTIONS FOR STEM-CELL RESEARCH.

SCIENTISTS IN CALIFORNIA REPORT THAT INJECTION OF HUMAN NEURAL STEM CELLS APPEARED TO REPAIR SPINAL CORDS IN PARTIALLY PARALYZED MICE, HELPING THEM TO WALK AGAIN.

RESEARCHERS AT THE SALK INSTITUTE FOR BIOLOGICAL STUDIES IN LA JOLLA, CALIFORNIA SUCCESSFULLY IMPLANTED UNDIFFERENTIATED HUMAN EMBRYONIC STEM CELLS INTO THE DEVELOPING BRAINS OF TWO-WEEK-OLD MOUSE EMBRYOS. AS THE ANIMALS GREW, THE CELLS MATURED INTO FULLY FUNCTIONAL ADULT BRAIN CELLS INTEGRATED WITH THE ANIMALS' NERVOUS SYSTEMS.

AUSTRIAN SCIENCE FUND RESEARCHERS DISCOVERED THAT THE PROTEIN SPARC AFFECTS THE ACTIVITY OF THE GENE RESPONSIBLE FOR THE EMERGENCE OF HEART CELLS FROM UNDIFFERENTIATED EMBRYONIC STEM CELLS. THIS FINDING MAY HELP IN DEVELOPING TREATMENT FOR HEART ATTACKS.

2006 GOVERNOR ROBERT EHRLICH SIGNS THE MARYLAND STEM CELL RESEARCH ACT, WHICH ALLOCATES $15 MILLION FOR EMBRYONIC STEM-CELL RESEARCH GRANTS.

THE SENATE CONSIDERS A BILL THAT EXPANDS FEDERAL FUNDING OF EMBRYONIC STEM-CELL RESEARCH. THE HOUSE PASSED ITS VERSION OF THE BILL IN 2005.

PRESIDENT BUSH VETOES THE BILL—THE FIRST USE OF HIS VETO POWER IN HIS PRESIDENCY.

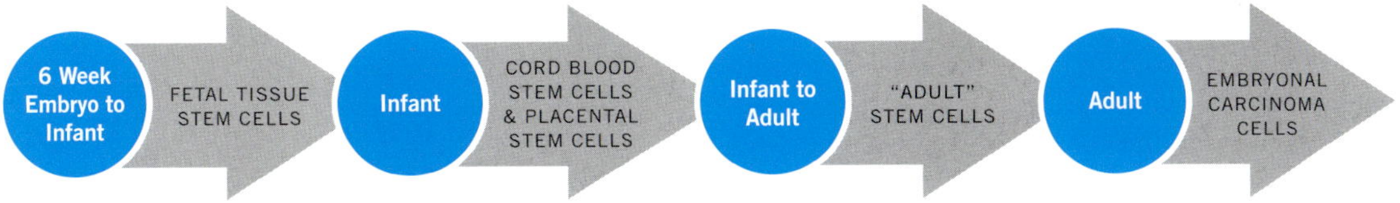

6 Week Embryo to Infant — FETAL TISSUE STEM CELLS

Infant — CORD BLOOD STEM CELLS & PLACENTAL STEM CELLS

Infant to Adult — "ADULT" STEM CELLS

Adult — EMBRYONAL CARCINOMA CELLS

2006 SCIENTISTS UNVEIL A NEW TECHNIQUE THEY CLAIM COULD BREAK THE POLITICAL DEADLOCK OVER HUMAN EMBRYONIC STEM CELLS.

RESEARCHERS WITH THE COMPANY ADVANCED CELL TECHNOLOGY SAY IT'S POSSIBLE TO REMOVE A CELL FROM AN EMBRYO WITHOUT HARMING THE EMBRYO AND THEN GROW THE CELL IN A LAB DISH. THAT SINGLE CELL COULD THEN BE USED TO DERIVE EMBRYONIC STEM CELLS.

MISSOURI VOTERS BACK A CONSTITUTIONAL AMENDMENT THAT SAFEGUARDS EMBRYONIC STEM-CELL RESEARCH IN THE STATE. MISSOURI'S LEGISLATURE HAD BEEN TRYING TO BAN SUCH RESEARCH IN THE STATE.

RESEARCHERS AT HARVARD UNIVERSITY ANNOUNCED PLANS TO CREATE NEW EMBRYONIC STEM-CELL LINES THROUGH SOMATIC-CELL NUCLEAR TRANSFER. IF SUCCESSFUL, THEIR EFFORT WOULD GENERATE STEM-CELL LINES PARTICULARLY WELL SUITED TO THE STUDY OF BLOOD DISEASES AND DIABETES. BECAUSE OF FEDERAL RESTRICTIONS, THEIR WORK IS PRIVATELY FUNDED.

RESEARCHERS GENERATE NEW LINES OF HUMAN EMBRYONIC STEM CELLS WITHOUT DESTROYING THE EMBRYO. WITH THIS DEVELOPMENT, SCIENTISTS HOPED TO CIRCUMVENT THE US BAN ON FEDERAL FUNDING OF STEM-CELL RESEARCH, PAVING THE WAY FOR MEDICAL ADVANCES.

SCIENTISTS IN THE UK CREATE THE FIRST-EVER ARTIFICIAL LIVER CELLS USING UMBILICAL-CORD-BLOOD STEM CELLS.

2007 JANUARY 7, RESEARCHERS AT WAKE FOREST UNIVERSITY AND HARVARD UNIVERSITY REPORT THAT STEM CELLS DRAWN FROM AMNIOTIC FLUID DONATED BY PREGNANT WOMEN HOLD MUCH THE SAME PROMISE AS EMBRYONIC STEM CELLS. THEY REPORTED THEY WERE ABLE TO EXTRACT THE STEM CELLS FROM THE FLUID, WHICH CUSHIONS BABIES IN THE WOMB, WITHOUT HARM TO MOTHER OR FETUS AND TURN THEIR DISCOVERY INTO SEVERAL DIFFERENT TISSUE CELL TYPES, INCLUDING BRAIN, LIVER, AND BONE.

FEBRUARY 28, IOWA GOVERNOR CHET CULVER EASES LIMITS ON STEM-CELL RESEARCH. THE LEGISLATION ALLOWS MEDICAL RESEARCHERS TO CREATE EMBRYONIC STEM CELLS THROUGH CLONING, BUT PROHIBITS REPRODUCTIVE CLONING OF HUMANS.

2007 MARCH 16, AFTER APPROVING NEARLY $45 MILLION FOR EMBRYONIC STEM-CELL RESEARCH IN FEBRUARY 2007, CALIFORNIA'S STEM-CELL AGENCY AUTHORIZES ANOTHER $75.7 MILLION TO FUND ESTABLISHED SCIENTISTS AT 12 NON-PROFIT AND ACADEMIC INSTITUTIONS.

APRIL 11, THE SENATE PASSES A BILL THAT WOULD EXPAND FEDERAL FUNDING FOR EMBRYONIC STEM-CELL RESEARCH. THE BILL PASSES 63-34, JUST SHY OF THE TWO-THIRDS MAJORITY NEEDED TO PROTECT THE LEGISLATION FROM PRESIDENT BUSH'S PROMISED VETO.

MAY 30, CALIFORNIA GOVERNOR ARNOLD SCHWARZENEGGER ANNOUNCES AN AGREEMENT BETWEEN THE UNIVERSITY OF CALIFORNIA AT BERKELEY AND CANADA'S INTERNATIONAL REGULOME CONSORTIUM TO COORDINATE STEM-CELL RESEARCH AT BOTH INSTITUTIONS. THE ONTARIO INSTITUTE OF CANCER RESEARCH DONATES THE FIRST $30 MILLION TO FUND A CANCER STEM CELL CONSORTIUM TO ADVANCE WORK ON POTENTIAL CANCER TREATMENTS.

JUNE 6, RESEARCHERS AT WHITEHEAD INSTITUTE IN MASSACHUSETTS MODIFY A SKIN CELL SO THAT IT BEHAVES LIKE AN EMBRYONIC STEM CELL, EASING SOME ETHICAL CONCERNS THAT CLONING EMBRYONIC STEM CELLS REQUIRES THE DESTRUCTION OF A HUMAN EMBRYO. AT HARVARD UNIVERSITY, SCIENTISTS MAKE IT POSSIBLE TO CLONE MICE FROM PREVIOUSLY FERTILIZED EGGS.

JUNE 20, PRESIDENT BUSH VETOES LEGISLATION THAT WOULD HAVE EASED RESTRAINTS ON STEM-CELL RESEARCH.

NOVEMBER 14, SCIENTISTS CLONE EMBRYOS FROM THE CELLS OF AN ADULT MONKEY AND DERIVE STEM CELLS FROM THOSE CLONED EMBRYOS.

NOVEMBER 20, TWO INDEPENDENT TEAMS OF SCIENTISTS REPORT ON A METHOD FOR MAKING HUMAN EMBRYONIC STEM CELLS WITHOUT DESTROYING A HUMAN EMBRYO. BY ADDING A COCKTAIL OF FOUR GENETIC FACTORS TO RUN-OF-THE-MILL HUMAN SKIN CELLS, TWO SCIENTIFIC TEAMS, ONE IN JAPAN AND ONE IN THE USA, ISOLATE CELLS THAT BEHAVE JUST LIKE EMBRYONIC STEM CELLS. THE RESEARCHERS CAUTION THERE ARE MANY STEPS BEFORE THESE CELLS ARE USEFUL FOR HUMAN THERAPIES. BUT THE WORK IS HAILED IN THE FIELD AS A CRITICAL STEP FORWARD, BOTH SCIENTIFICALLY AND ETHICALLY.

2009 MARCH 27, PRESIDENT OBAMA SUPPORTS FEDERAL FUNDING FOR EMBRYONIC STEM-CELL RESEARCH.

A Global Phenomenon /

The race to realize the potential of stem cells is nothing short of a worldwide medical revolution, and the economic ramifications are immense. The global stem cells market is projected to top $32 billion by 2012, according to a 2008 report by Research And Markets, an international market-research firm.[10] Many nations are pinning key economic development plans—and their scientific leadership positions—on the future of these tiny cells.

Australia, China, India, Japan, Singapore, and South Korea "all see stem-cell research as a way to get ahead in biotech," reported *Business Week*.[11] "Some governments have focused on importing talent. China, for instance, has recruited scientists from top universities in the US to run research centers on the mainland." But the political winds keep changing, as evidenced by California's passing of Proposition 71, which is providing $300 million a year to scientists conducting such research in the state, and similar laws have been passed in several other states. Those initiatives make it harder for Asian countries to attract top scientists, *Business Week* added. More funding for various types of stem-cell research will become available as the science evolves.

In recent years, many nations have committed vast resources to advancing stem cell research. Some of the major initiatives include the following:

- The UN Millennium Project, established to better understand animal development and regeneration, launched The RIKEN Center for Developmental Biology in 2000. The center, located in a world-class biomedical research park in Kobe, Japan, enjoys the support of local and national governments as well as public, academic, and corporate research organizations.
- Scotland has emerged as a leader in stem-cell research. The nation boasts a powerful life science research and technology base of more that 500 organizations and more that 26,000 employees. Multinationals based there include Stem Cell Sciences Ltd, CXR Biosciences, Invitrogen, Angel Biotechnology, and Geron Corporation. Edinburgh is home to Europe's most robust stem-cell research community with two key centers of excellence: The Centre for Regenerative Medicine, headed by Sir Ian Wilmut (creator of Dolly, the cloned sheep); and The Institute for Stem Cell Research, the lead contractor for the European Union's $19-million FP6 (Sixth Framework Programme) project to create a European consortium for stem-cell research.
- Singapore, which in 2004 opened Biopolis, a 2-million-square-foot complex devoted to stem-cell research, also has its sights set on leadership in the emerging science. And like many other nations, its outlook has been bolstered by Washington, D.C. "Bush administration policies that restrict federal money for stem-cell research have prompted an increasing number of top scientists to pack their bags and head for this equatorial city," reported *The New York Times* in August 2006.[12]

- In 2004 Indian president APJ Abdul Kalam identified stem cell research as a national priority. That same year India began drawing plans for a stem-cell initiative, with a special focus on clinical applications of stem cells in ophthalmology, cardiology, and spinal cord repair.[13] "A key objective of the initiative is to promote 'stem cell city clusters'," reported the London-based international Science and Development Network in 2005. "These would link all publicly and privately funded research groups within a city, enabling them to share facilities, ideas, and research and business opportunities, as well as promoting interactions between researchers and clinicians."
- India, an emerging key player in the computational and systems biology (bio-IT) arena, in 2005 created the country's first stem cell translational research center at the Christian Medical College, Vellore, in the state of Tamil Nadu. In 2006, India's first stem-cell transplant center opened in Chennai, the capital of Tamil Nadu. LifeCell, an Indian stem-cell banking and research company, launched the center with US-based Cryo-Cell Inc.
- In 2006, an international jury awarded a film about stem cells the top prize at the European science media festival. Scientists working for EuroStemCell, a research consortium based at Edinburgh University, produced the fascinating 15-minute documentary (available on the Internet) entitled *A Stem Cell Story*.
- In 2006, the South Korean government announced that it would spend $454 million over the next 10 years in a quest to become one of the top three global leaders in stem cell research.[14] The plan includes developing a stem-cell-research infrastructure and attracting more scientists.
- Two of the US's most prominent cancer researchers, Neal G Copeland and Nancy A Jenkins, relocated to the tiny island-state of Singapore in 2006 to work at the Institute of Molecular and Cell Biology. "The husband-and-wife team, who worked for 20 years at the National Cancer Institute in Maryland, said politics and budget cuts had left financing in the United States too hard to come by," said *The New York Times.* Copeland and Jenkins are just two of many leading US researchers who have moved to Singapore in recent years in response to its biomedical freedom and generous funding.

Funding /

The potential of stem cells inspires great hope, but there is far more to learn. **Researchers want to know how cells maintain the never-ending self-renewal that sustains the organism as a whole.** To learn this, they must understand two essential processes—the need for old cells destined for replacement to be able to disengage from their environmental milieux, and the requirement for preparing new cells in replacement.

In the span of about 10 years, stem cell research has grown from infancy to a key area of research globally, fueled mainly by government funding. Unfortunately, the time required for the development of much-needed clinical treatments is beyond that usually accepted by venture-capital investors, but private funding will increase as treatments hit the marketplace. At some stage, private money will likely match public investment in the US, the UK, Singapore, Israel, and other countries interested in stem cell research.

US scientists believe that rapid progress in this research depends largely upon federal funds. Federal dollars help attract the top scientists and ensure that new discoveries are widely shared among research facilities, and that the research is directed toward the greatest public good.

06.6 / A RESEARCHER FOCUSES ON SELECTING COLONIES OF CULTURED HUMAN EMBRYONIC CELLS.

Biocontainment /

Introduction /

Since the terrorism attacks on September 11, 2001, the US has made biocontainment research a top priority. At the expense of other areas of research, over 50 percent of US federal funding was spent on biodefense during the Bush administration. That is now changing under the Obama presidency, with less spending being directed toward NIH and "green research."

Biocontainment research is undertaken to respond to two types of catastrophes, infectious disease outbreaks and acts of terrorism. Since the 1950s, scientists have studied the containment of infectious diseases, which can be used in terrorism acts. Bioterrorism and infectious disease cannot be contained by national borders. Hazardous substances and infectious disease can travel quickly by airplane, ship or other means, so an uncontained local outbreak puts the rest of the world at risk. Because developing countries lack the infrastructure to contain mass outbreaks, the global community must be counted on to mobilize a highly coordinated response. In a recent success story, researchers, agencies, and nations collaborated in 2004–05 to minimize the global spread of severe acute respiratory syndrome (SARS).

An epidemic is defined as spreading beyond a local population; a pandemic reaches worldwide proportions. The following is a historical summary of major infectious-disease outbreaks by time period and estimated deaths:

Years	Infectious Diseases	Estimated Deaths
430 BC	TYPHOID FEVER	25 PERCENT OF ATHENS, GREECE
165–180	ANTONINE PLAGUE	5 MILLION
541–542	BUBONIC PLAGUE	25 PERCENT OF MEDITERRANEAN
14TH CENTURY	BLACK DEATH	75 MILLION
1855	PANDEMIC IN ASIA	12 MILLION
1918–20	SPANISH FLU	50 MILLION
1957–58	ASIAN FLU	2 MILLION
1968	HONG KONG FLU	750,000
2004–05	SARS	774

In contrast with the Spanish Flu, which killed about 50 million people, 15.6 million military personnel and civilians died in the First World War during this same time period. And during Napoleon's retreat from Russia, more of his soldiers died from typhus than in battle. Throughout history, more people have died from infectious disease than by war. Thanks to advancements in science, however, that is not currently the case. A more disturbing take on that development, however, is that even in these modern times, warfare is rampant. Furthermore, an uncontrolled pandemic caused by bird flu, for example, could reverse that trend.

Among infectious diseases, AIDS, malaria, and neglected tropical diseases (NDTs) are among the most deadly today. AIDS has led to the deaths of more than 22 million people since it was first recognized in 1981. In recent years, it has killed millions of people in the poorest countries. In 2007 alone, 2.1 million people died from AIDS-related illnesses and 2.5 million people became newly infected, reported WHO.[15] Like malaria, AIDS can be prevented, but just one in five people at risk of HIV has access to the information and tools they need to prevent it. Furthermore, millions are in urgent need of antiretroviral medicines.

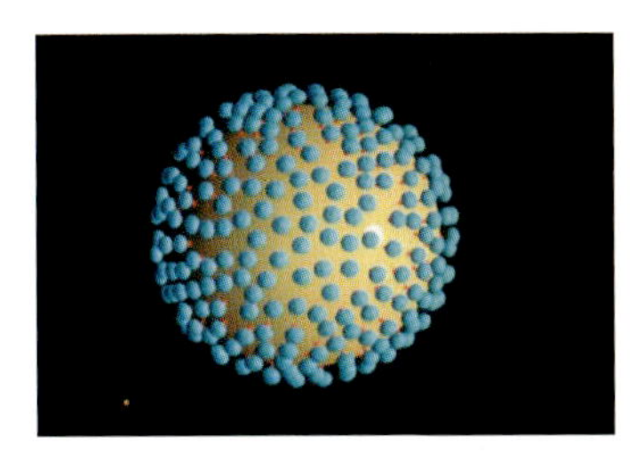

06.7 / THE AIDS VIRUS

Malaria, one of the greatest killers in human history, still kills about 1.2 million people a year, mostly African children under the age of five, reports WHO.[16] This is unacceptable because malaria is preventable. After the Second World War, malaria disappeared almost entirely due to the use of DDT. However, with the banning of this pesticide, malaria is once again a serious problem. In 2006, WHO advocated reintroducing the use of DDT and other insecticides under restricted conditions, such as indoor spraying, to control malaria throughout Africa.

"The scientific and programmatic evidence clearly supports this reassessment," reported WHO. In 2007, however, WHO somewhat altered its position to more closely align with the Stockholm Convention, saying that while WHO still recommends the limited use of DDT, it is "very much concerned with health consequences [and] is committed to making sure alternatives [to DDT] are soon available."[17] Humanitarians have also clearly demonstrated that insecticide-treated bed nets, which cost just under $5 apiece and last for five years, can break the transmission of lymphatic filariasis and greatly reduce the transmission of malaria.

Neglected tropical diseases kill and maim millions of people a year, but they are less well known than AIDS or SARS because they principally afflict poor people residing in the tropics. Of the 13 NTDs, nine of them (the seven helminth infections, plus leprosy and trachoma) can be easily and cost-effectively prevented or treated, reported economist Jeffrey Sachs in *Scientific American*. "As President Jimmy Carter has shown through his steadfast personal leadership over 20 years, filtering water through cheesecloth can dramatically reduce the burden of dracunculiasis," said Sachs.[18]

In response to NTDs, pharmaceutical companies Merck & Co., GlaxoSmithKline, Johnson & Johnson, Pfizer, Novartis, and Sanofi-Pasteur have donated medicines and other resources. But, argues Sachs, governments must do more: "The US has recently committed $15 million to fight against NTDs—a start but still less than one tenth of the $250 million or so a year needed for a comprehensive campaign for Africa."

Sachs estimates that "comprehensive, Africa-wide control of malaria and NTDs together would probably cost no more than $3 billion a year ... If each of the billion people in the rich world devoted the equivalent of one $3 cup of coffee a year to the cause, several million children every year would be spared death and debility."

The last two pandemics in the 20th century (the 1957 Asian Flu and the 1968 Hong Kong Flu) involved bird viruses, which can mutate to easily infect humans, who have no immunity. The current strain of bird flu, called Avian Influenza H5N1, was discovered in 2003, and by January 2008 it had killed 221 of the 353 people who had contracted it.[19] The most deaths, 98, had occurred in Indonesia.

According to a 2005 report from the public health advocacy group Trust for America's Health, if only a moderately severe strain of a bird flu pandemic erupts in the US, over 500,000 Americans could die and over 2.3 million could require hospitalization.[20] The nonprofit health watchdog projects that 66.9 million Americans would be at risk of contracting such a disease. And if as projected, more than 2 million required hospitalization, most would be out of luck. In 2005 there were some 965,256 staffed hospital beds in the US.

Humans diseases contracted from animals are referred to as zoonotic diseases. In the past 100 years, an average of one such disease every 10 years has been eradicated from animals. When a highly infectious disease in an animal is discovered, typically all animals in that area are destroyed, usually costing the agriculture industry billions of dollars.

Zoonoitic diseases discovered during the past 25 years include the following:

Year	Disease
1980s	LYME DISEASE
1981	RABBIT HAEMORRHAGIC DISEASE VIRUS
1985	BOVINE SPONGIFORM ENCEPHALOPATHY
1987	PORCINE REPRODUCTIVE AND RESPIRATORY SYNDROME
1988	SEAL MORBILLIVIRUS
1990s	TURKEY RHINOTRACHEITIS
1993	ESCHERICHIA COLI 0157:H7
1994	EQUINE HENDRA VIRUS
1994	CANINE DISTEMPER IN LIONS
1994	TAURA SYNDROME VIRUS IN SHRIMP
1995	SALMONELLA DT104
1997	PFIESTERIA SP.
1999	NIPAH VIRUS

The above list does not include non-zoononic diseases, such as foot-and-mouth disease, that do not cause disease in humans. Like zoononic diseases, however, non-zoononic diseases typically cost the agriculture industry dearly due to the number of animals that must be destroyed.

Time and again, science and commitment have proven worthy adversaries against deadly diseases. The discovery of the polio vaccine in 1955 has eradicated that widespread killer from everywhere in the world except in remote areas where the vaccine is not used. And, when the SARS outbreak occurred a few years ago, it was mostly contained in Asia. The few cases in the US were addressed quickly, and countries around the world were networked to minimize the casualties. (SARS was relatively easily contained because it is not infectious during the initial few days of fever. A major influenza, on the other hand, would be more deadly because it is infectious prior to the onset of fever.)

06.8 / THE SWINE-FLU VIRUS.

06.9 / MALARIA, A MOSQUITO-BORNE INFECTIOUS DISEASE, KILLS ABOUT 1.2 MILLION PEOPLE EACH YEAR.

Beyond infectious disease, mankind's other great killer is war. In all major wars in which the US has been involved, nearly 95 million people have perished:

US Conflict	Estimated Deaths
VIETNAM WAR	4.8 M
KOREAN WAR	1.8 M
SECOND WORLD WAR	72 M
FIRST WORLD WAR	15.6 M
AMERICAN CIVIL WAR	620,000
SPANISH AMERICAN WAR	3,357
REVOLUTIONARY WAR	37,980

These deaths do not include the other wars that have taken place in the world over the past 200 years. Nor do they include democide—genocide, politicide, or mass murder perpetrated by government. Those numbers are staggering and disturbing. Approximately 180 million people have been murdered by democide in the past 200 years, an average of almost a million people each year.

The US pays a hefty price for war. The following are estimated US costs, in 2007 dollars, for recent wars:

US Conflict	Estimated Cost (Billions)
IRAQ AND AFGHANISTAN	$764
VIETNAM WAR	$600
KOREAN WAR	$430
SECOND WORLD WAR	$3,500
FIRST WORLD WAR	$2,300

As stated in Chapter 3, the average total cost to develop a successful new drug is approximately $800 million. Even at an average cost of $1 billion per successful drug discovery, **the US cost of the war in Iraq and Afghanistan (as of 2007) potentially could have resulted in 764 significant research discoveries to improve and extend the lives of billions of people worldwide.**

Opportunities /

Advances in science and technology have greatly aided in the battle against infectious diseases. Researchers can now study a new virus as it mutates and then compare notes and come to a consensus on how to respond. Furthermore, satellite phones, the Internet, and e-mail enable public health workers to coordinate response initiatives and transmit urgent messages from remote villages. Global surveillance and collaboration is imperative to minimize the impact of infectious diseases and possible bioweapons.

Among its other missions, WHO seeks to protect world citizens against deadly outbreaks or bioterrorism attacks. Several epidemic diseases within the scope of the global health organization have been associated with biological warfare. In recent years, WHO has improved epidemiology surveillance programs, response measures, and laboratory facilities. In addition to analyzing and disseminating information on outbreaks, the organization administers the International Health Regulations, a global and politically neutral framework within which surveillance and response networks can operate in a timely and coordinated manner.

The International Programme on Chemical Safety (IPCS), a joint venture of the United Nations Environment Programme, the International Labour Organization and WHO, evaluates the effects of chemicals on humans and the environment. It also sets guidelines on preparedness for and response to chemical incidents. The IPCS supports national chemical safety programs and 24-hour chemical information centers. It supports an electronically linked network of about 120 centers in 70 countries, allowing rapid access to toxicological, analytical, and clinical expertise.[21]

As stated, the US is committed to biocontainment research, which aims to prevent or quickly limit the national or international impact of a mass infectious disease or bioweapon. Project BioShield, implemented under the Bush administration, was designed to provide Americans with fast-response medical countermeasures against chemical, biological, radiological, or nuclear (CBRN) attacks. The fiscal year 2004 appropriation for the Department of Homeland Security included $5.6 billion over 10 years for the purchase of next-generation countermeasures, including vaccines or drugs, against anthrax and smallpox as well as other CBRN agents.

Several federal health agencies are accelerating basic microbial research and developing new ways to diagnose, prevent or treat infections that may be caused by the intentional release of a pathogen. According to the National Institute of Allergy and Infections Diseases (NIAID), "microbial genomics is a vital part of a comprehensive approach to biodefense. The genomes of several smallpox virus strains are already known, as are those for different hemorrhagic fever viruses." NIAID funds the genome sequencing of many other potential bioterrorism microbes and collaborates with the Defense Advanced Research Products Agency to sequence potential agents of bioterror such as the bacteria that cause brucellosis, Q fever, gangrene, and epidemic typhus. In 2002, the genome sequences of the anthrax and plague bacteria were completed.[22]

The genetic blueprints revealed by genome sequencing will identify key genes required for microbes to infect people and thrive. According to NIAID, "The protein molecules encoded by some of those genes will be likely targets for new drugs and vaccines to protect the public from disease."

06.10

06.11

Funding /

Since 2001, the US has spent massive resources on preparing the nation against a bioterrorist attack. The 2008 federal budget allocated a record-breaking $142 billion for research and development. This included an increase of $309 million for civilian biodefense programs over fiscal year 2007. US biodefense funding has increased from $576 million in 2001 to an estimated $5.1 billion in 2007. The fiscal year 2008 budget called for $5.4 billion to be allocated to biodefense.[23] Not surprisingly, to increase chances for funding, many academic grants are tied to bioterrorism.

Public Health Laboratories /

Since 2001, US public health systems are much better prepared for bioterrorism and other incidents that threaten mass casualties. Public health labs have been coordinated at local, regional, and national levels. By 2004, all 50 states had systems in place to rapidly detect and respond to bioterrorism attacks, including mass vaccination plans. Also by 2004, 90 percent of CDC awardees had the ability to initiate a field investigation within six hours of receiving an urgent disease report, according to the Department of Health and Human Resources.[24]

The flow of samples in a public health laboratory is critical in biosafety and biosecurity. Samples should be processed into the facility, and be separated from the public, clerical personnel, and laboratory personnel who are not required to interact with the samples. Specimens should have a clear flow through the facility from entry to exit.

Samples brought into the lab must be cataloged, where each sample is uniquely identified and logged in. Packages containing potentially hazardous biological samples should be opened in a biological safety cabinet or fume hood to protect laboratory personnel in the event of a damaged or leaking shipping container. Samples are then distributed to the appropriate laboratory areas for processing. Sample transfer should minimize the potential for contact with non-laboratory areas.

It is apparent that the inclusion of bioterrorism agents will pose challenges to the public-health laboratory infrastructure. Handling these agents in a manner that maintains the health of laboratory workers will be critical to responding successfully to bioterrorism incidences. It is also important not to overreact by creating complex laboratories and protocols that will impede rapid and accurate diagnosis.

A network of laboratories has been established within the public health sector to ensure preparedness for certain diagnostic capacities at various levels. It is envisioned that routine diagnostic work will be conducted in BSL-2 laboratories (designated as A labs). When the level of suspicion goes up that there may be an agent involved that requires higher containment or more sophisticated diagnostic techniques, the materials are sent to B labs (BSL-2 facilities with BSL-3 work practices and procedures). C labs are those that function in BSL-3 facilities with full BSL-3 practices and procedures. D labs can perform all the lower-level tests and have the unique capacity to perform work at enhanced BSL-3 or BSL-4. These labs will also serve as the repositories for all samples associated with bioterrorist events.

There is still much to be done to prepare any country to effectively address public-health disasters. This was very clear when Hurricane Katrina hit New Orleans and the surrounding areas. The number of natural disasters has increased significantly in the past 25 years. Much of this is attributed to global warming. Continuous developments and international collaboration are needed to effectively address bioterrorism, infectious diseases, and natural disasters. This is where global research can play a major role. There are several private companies and non-profit foundations that can support this effort, including the Gates Global Health Initiative, and Warren Buffett.

06.10 / THE DEPARTMENT OF HOMELAND SECURITY'S BIOCONTAINMENT RESEARCH FACILITY (NBACC) AT FORT DETRICK, FREDERICK, MARYLAND, WAS OPENED IN 2009.

06.11 / THE NATIONAL BIODEFENSE ANALYSIS AND COUNTERMEASURES CENTER (NBACC) AT FORT DETRICK, FREDERICK, MARYLAND.

06.12 / THE GALVESTON NATIONAL LABORATORY IS A PART OF THE NATIONAL PUBLIC HEALTH SYSTEM IN THE USA AND SUPPORTS INTERNATIONAL RESEARCH AS WELL.

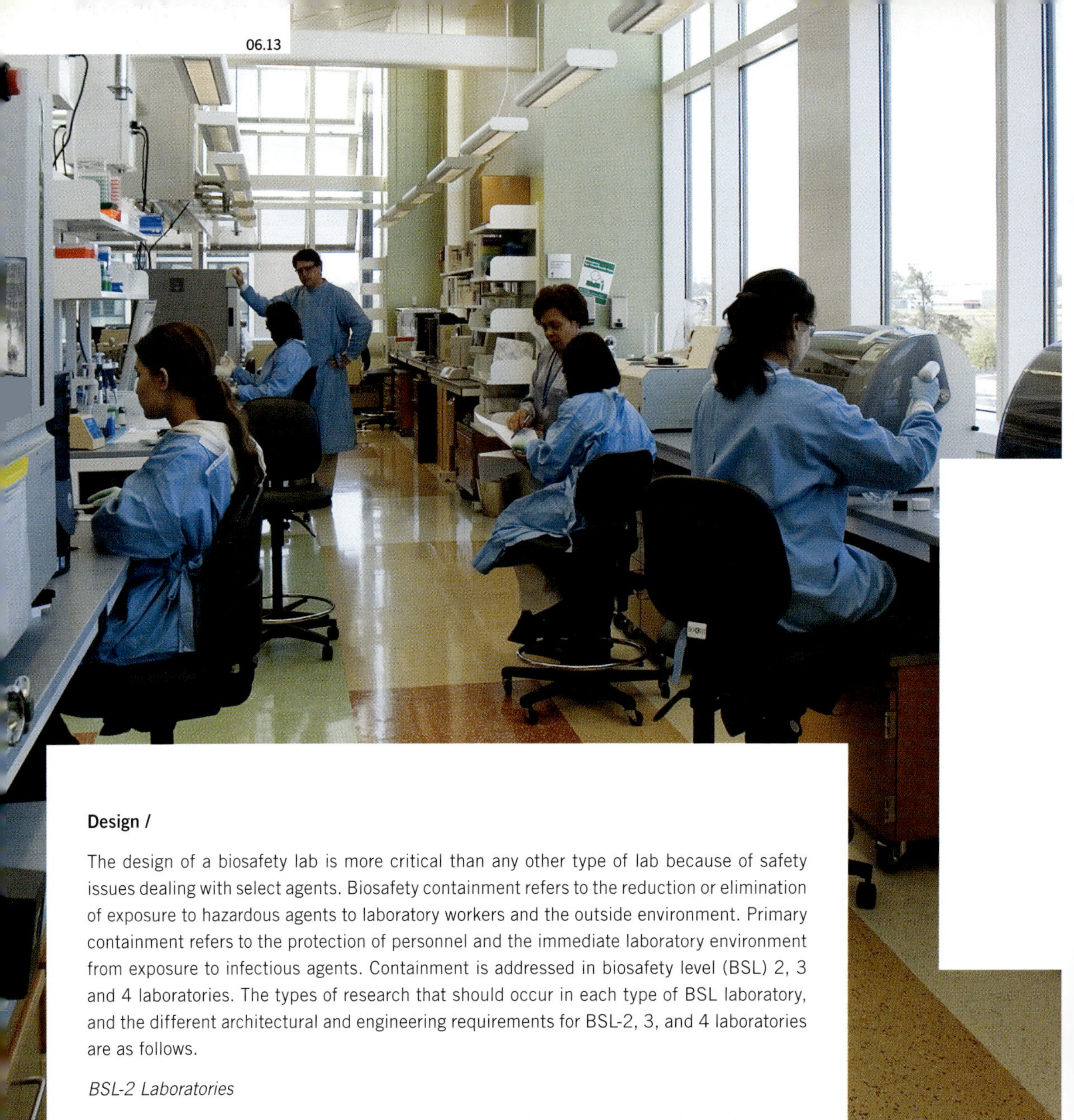

Design /

The design of a biosafety lab is more critical than any other type of lab because of safety issues dealing with select agents. Biosafety containment refers to the reduction or elimination of exposure to hazardous agents to laboratory workers and the outside environment. Primary containment refers to the protection of personnel and the immediate laboratory environment from exposure to infectious agents. Containment is addressed in biosafety level (BSL) 2, 3 and 4 laboratories. The types of research that should occur in each type of BSL laboratory, and the different architectural and engineering requirements for BSL-2, 3, and 4 laboratories are as follows.

BSL-2 Laboratories

BSL-2 laboratories are suitable for work with agents of a potentially moderate hazard to personnel and the environment. Personnel are trained in hazard identification and work procedures; access to the work area is controlled and limited; hazard signs are posted; extreme precautions with sharps are observed; and special equipment may be used to contain or control chemical fumes, splatters, or biological aerosols. Emphasis is on safe practices and procedures. Restrictions on smoking, eating, and drinking in the laboratory areas are enforced to reduce ingestion potential. Gloves, gowns, and aprons are worn.

Other personal protective equipment is worn as needed, for example to protect mucous membranes. Hand washing after removal of gloves and before leaving the work area is required. Instruments and work surfaces are decontaminated and cleaned. Waste is decontaminated or processed for incineration. Samples are labeled, including hazard warnings, and contained for transport to other locations.

06.14

It is standard policy to have the lab supervised by an appropriately qualified manager. Immunizations are offered and medical services are available to all staff members. Standard operating procedures are developed and followed by everyone. Facility requirements dictate that the laboratories are away from public areas and doors are lockable. Consideration is given for directional inward airflow without re-circulation to other areas.

BSL-3 Laboratories

The primary focus is on containing the potential exposure to pathogens spread by aerosol and to infectious diseases such as M. tuberculosis, St. Louis encephalitis virus, and Coxiella burnetii. The design criteria for BSL-3 laboratories starts with meeting the BSL-2 criteria, then including a separate building or isolated zone, double door entry, directional inward airflow, and single-pass air. Also required are enclosures for aerosol-generating equipment, sealed rooms, and water-resistant walls, floors, and ceilings for easy cleaning.

BSL-3 laboratories are suitable for work with agents that carry potentially lethal diseases and can be transmitted by way of inhalation. Personnel receive specific training in handling materials infected with these specific agents. Supervision is provided by competent scientists who are experienced in working with such agents. Localized containment or ventilation devices are used to control or contain fumes, splatters, and biological aerosols. Special engineering controls and appropriate personal protective clothing and equipment, including respirators, are mandatory.

In addition to the principles of the BSL-2 lab, additional medical surveillance procedures might be applicable for periodic TB skin testing, and serum collection and testing. As well, biohazard warning signs indicating suspect agents and necessary precautions are posted.

06.13 / A CENTERS FOR DISEASE CONTROL AND PREVENTION BSL-2 LABORATORY.

06.14 / A BSL-3 LABORATORY.

06.15

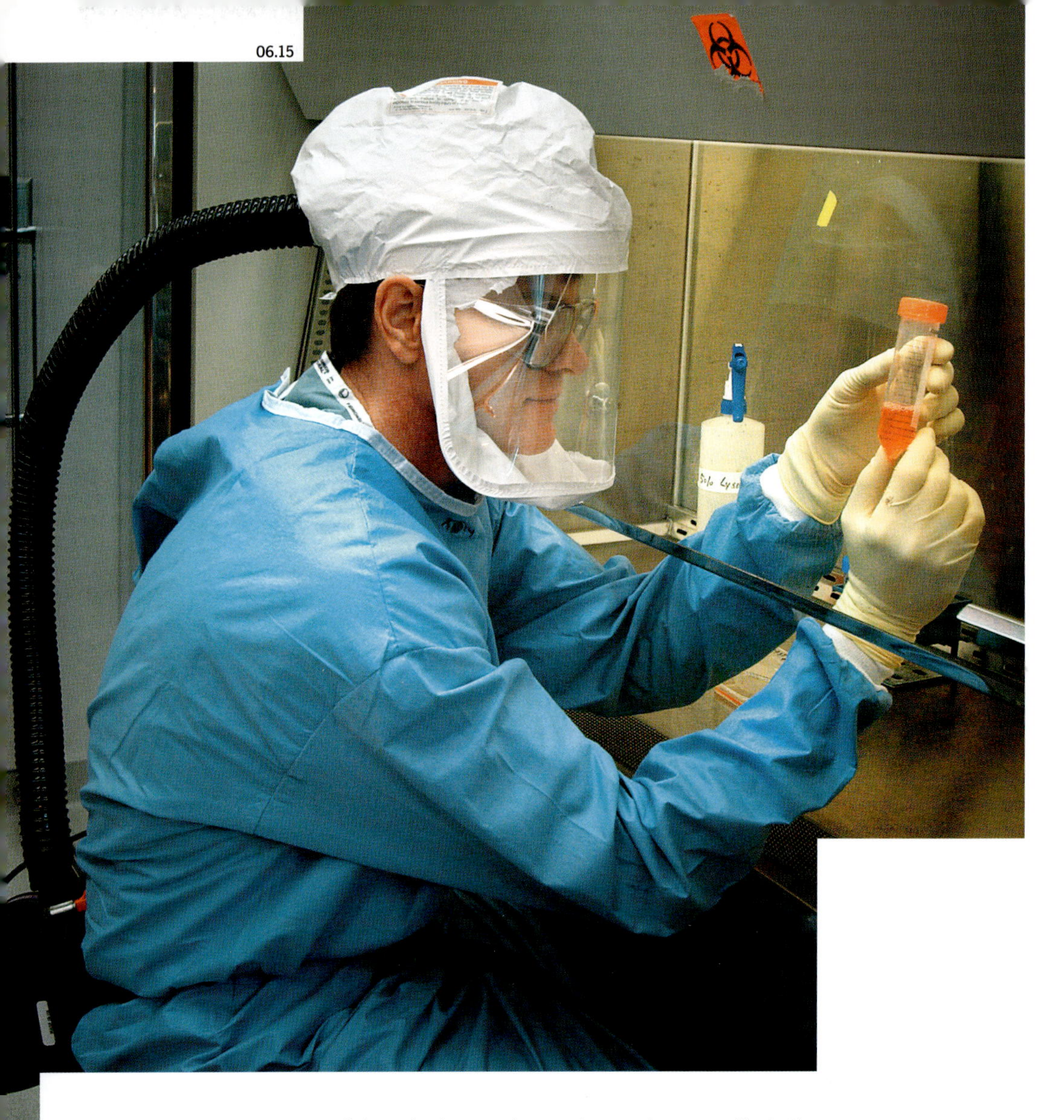

All personnel demonstrate proficiency in the practices and procedures specific to the nature of the hazard. Many researchers in biocontainment labs prefer direct access between the BSL-2 and BSL-3 areas. BSL-3 laboratories are a key growth area in construction for government, private, and academic sectors worldwide.

BSL-4 laboratories

BSL-4 laboratories have seen a construction boom in the US in the past five years, as well as a modest increase around the world. BSL-4 labs are required when agents, specifically viruses, pose a high individual risk of aerosol-transmitted laboratory infection and life-threatening disease and there is no specific therapy or vaccine. This type of containment laboratory must be designed and constructed to specific containment requirements to minimize the potential for personnel exposure and to prevent dissemination of BSL-4 organisms into the environment. Personnel enter and leave only through the clothing-change and shower rooms. Personnel must wear a one-piece positive-pressure suit ventilated with a life-support system, and shower before they leave the facility. Emergency backup systems must be at 100 percent. All exhaust

06.16

In the 1970s, Karl M. Johnson led a small research team to a remote area of Bolivia where the Ebola virus struck. Equipment was modest: four fume hoods were created out of plywood to create a makeshift containment lab staged off a trailer. Today, we have sophisticated, safe, portable biocontainment labs. However, these labs are not typically available in remote areas of developing countries where outbreaks are likely to occur due to lack of good healthcare facilities and practices.

air must go through high performance filters, and the rooms and equipment must be decontaminated. Waste must be cooked, killed, and monitored for the highest safety standards.

For animal research, vivarium space is also usually built to support BSL-2, 3, and 4 laboratories. By 2010, the US expects that each health department will have access to rapid, high-quality laboratory testing through a National Laboratory System of specimen collection, transport, testing, confirmation, and reporting.

06.15 / PERSONAL PROTECTIVE CLOTHING AND EQUIPMENT ARE MANDATORY IN BSL-3 LABORATORIES.

06.16 / A RESEARCHER IN A "SPACE SUIT" IN THE BSL-4 LAB AT THE GALVESTON NATIONAL LABORATORY. PERSONAL PROTECTIVE CLOTHING AND EQUIPMENT ARE MANDATORY IN BSL-3 LABORATORIES. THE FACILITY STAYED FULLY OPERATIONAL DURING HURRICANE IKE IN 2008.

Commentary /

The following are thoughts from Karl M. Johnson, MD, an internationally acclaimed virologist and an expert in biocontainment. Karl led the task force that investigated the first outbreak of the Ebola virus in Zaire in 1976. He made the first isolation of the virus, naming it after a river in the affected region.

Much of Karl's research has involved working in primitive areas under difficult conditions. While doing fieldwork, he contracted the Machupo virus, from which he

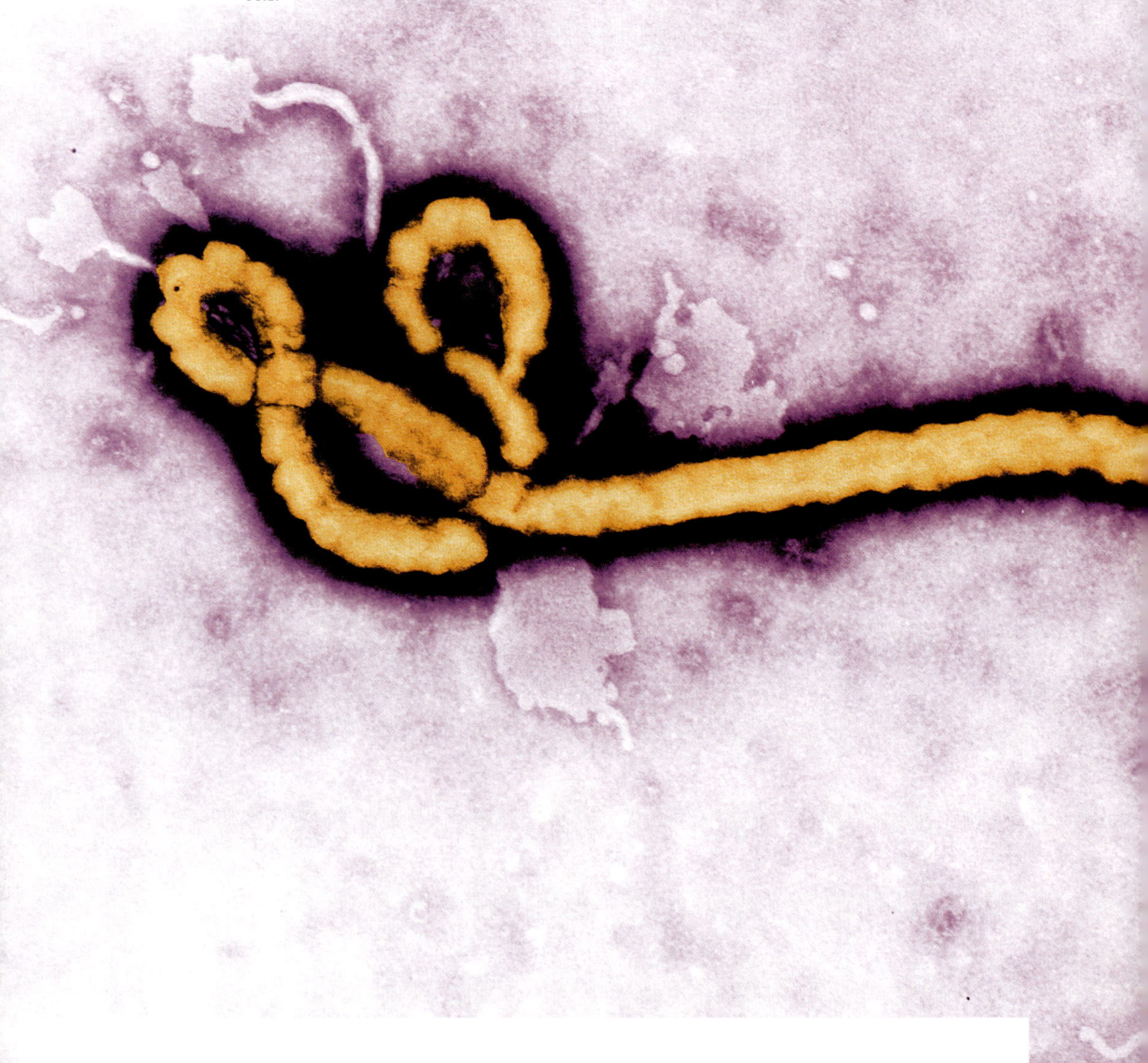

recovered. Karl's research has been written about in books such as *Virus Hunter* and *The Hot Zone*. From 1975 to 1981, Karl served as chief, Special Pathogens Branch, Virology Division at the Centers for Disease Control. During this era, he established the first space-suit BSL-4 laboratory, located in Atlanta, and a field research station in Sierra Leone. The Sierra Leone work defined the clinical and laboratory parameters of Lassa fever virus and led to the treatment of this deadly disease with the antiviral drug ribavirin.

Over the past seven years I have had the privilege to work with Karl on designing the Galveston National Laboratory and two biocontainment facilities for the Department of Homeland Security. The following are some of Karl's thoughts on the future of biocontainment laboratories.

"I do not see any significant changes in concepts for design or construction of (BSL-4) facilities in North America. Actually I think it unlikely that any more will be constructed in our country for the next 20 years. We will have major problems "digesting" those now about to come on line. Even if a terrorist attack utilizing one of the agents should occur, I believe the response will be limited to better support for those we have.

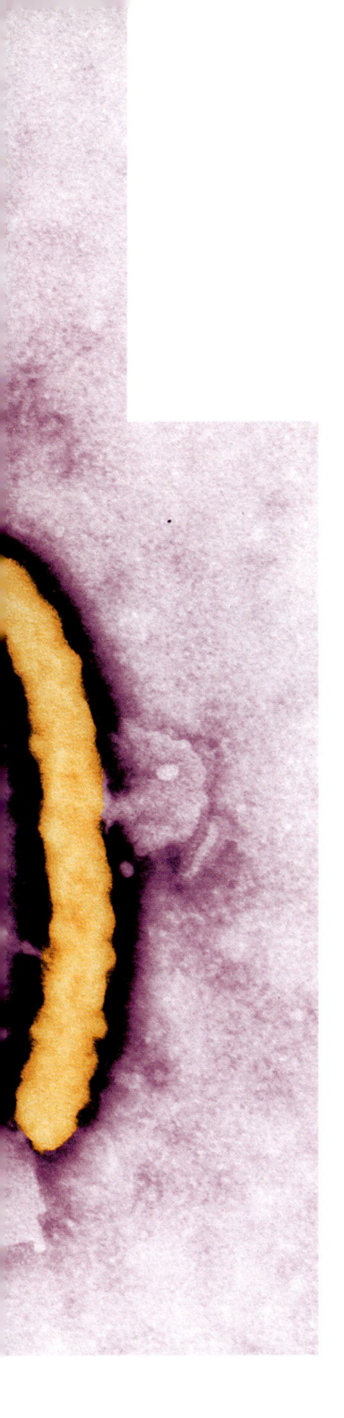

It is important to remember, also, that the viruses that need this level of containment are relatively few in number, do not produce disease in the US, with the exception of Hantavirus, and are not difficult to work with in either the molecular or macrological sense. I can see the day when vaccines are available for all of them, and when that happens it will be safe to continue work at BSL-3, as has now been accepted for Junin virus from Argentina.

The decision that the worst influenza viruses do not require Level-4 containment is further reason to suspect that more such labs will not be built here during my lifetime.

Perhaps the major change in lab configuration that might occur is to get rid of the external breathing air system and the hoses hanging from ceilings. I believe that development of such self-contained suits is underway in France. Before I would wear one, however, I would require new measurements of viral aerosols generated during laboratory accidents like broken tubes in centrifuges, and by animals infected with some of the more pathogenic agents.

I believe that future laboratories will occur in countries outside North America, but that they will be few and far between in time. There could be major improvements in construction of these if some careful mockup work were done. I think especially of materials used for floors, walls, and ceilings. We no longer have a requirement for the infamous pressure decay test in the US although our 21st-century edifices will be built to that standard. If I had the money I would experiment with much lighter materials and spend significant effort on modern paint and plastics to achieve good seals.

Will we see the day when robots do some of the repetitive work in BSL-4? I am not sure because I am no longer that close to machinery already in use for molecular manipulations, but I suspect that this, in fact, might be very useful. The NIH lab in Hamilton (Montana) might be the best candidate for some applications, as the intramural program will always have more ability to fund things than any academic institution, and is more likely to think outside the box than either DOD or DHS."

06.17 / THE EBOLA VIRUS

07. NEW TECHNOLOGY

Nanotechnology /

Science /

Described as a "new industrial revolution," nanotechnologies have the potential to change our lives dramatically. **Key areas that can expect to benefit from nanotechnology include the environment, communication, health, and production.** Through nanotechnologies, production processes may become cleaner, safer, and more competitive, and products may become smarter, more durable, and more user-friendly. Furthermore, nanotechnologies may provide innovative answers to the triple challenge of sustainable development: how to fuel economic growth while preserving the environment, and enhancing safety, security, and quality of life.[1]

The potential applications for nanotechnology are vast. The new science is expected to produce materials that are lighter than steel but 10 times stronger; sugar-cube sized digital storage units that can hold more information than the Library of Congress; and micro medical devices that are capable of detecting individual cancer cells and targeting them with specialized treatment.

Growth of aligned arrays of nanowires is of great importance in nanobiotechnology, because they are a fundamental structure for biosensing, manipulation of cells, electron field emission, and converting mechanical energy (such as body movement, muscle contraction, heart beating, and blood flow) into electricity to power nanodevices. Nanotechnology is the ability to control or manipulate on the atomic scale. Nanotechnology research and development is also at the molecular or macromolecular levels, in the length scale of approximately 1–100 nanometer range. This type of research involves creating and using structures, devices, and systems that have novel properties and functions because of their small size.

Tools developed through nanotechnology may be able to detect disease in a very small amount of cells or tissue. They may also be able to enter and monitor cells within a living body. "A nanometer is a sheet of paper that is about 100,000 nanometers thick ... almost as wide as a DNA molecule and 10 times the diameter of a hydrogen atom. It's about how much your fingernails grow each second and how far the San Andreas Fault slips in half a second. It's the thickness of a drop of water spread over a square meter. It's one-tenth the thickness of the metal film on your tinted sunglasses or your potato chip bag. The smallest lithographic feature on a Pentium computer chip is about 100 nanometers," Thomas Theis told a symposium at Sanford University.[2]

According to the National Cancer Institute, "Most animal cells are 10,000 to 20,000 nanometers in diameter. This means that nanoscale devices (having at least one dimension less than 100 nanometers) can enter cells and (allow) the organelles inside them to interact with DNA and proteins."[3]

Miniaturization will allow the tools for many different tests to be situated together on the same small device. This means that nanotechnology could make it possible to run many diagnostic tests simultaneously as well as with more sensitivity. **In general, nanotechnology may offer a faster and more efficient means for scientists to do much of what they do now.** Nanotechnology requires interdisciplinary research from at least two to three of the following areas: physics, chemistry, biology, material sciences, computer sciences, mathematics, and engineering.

There are two different approaches to nanotechnology—top-down and bottom-up. "Top-down" nanotechnology features the use of micro- and nano-lithography and etching. Here, small features are made by starting with larger materials (e.g. semi-conductors) and patterning and "carving down" to make nanoscale structures in precise patterns. Complex structures including microprocessors containing hundreds of millions of precisely positioned nanostructures can be fabricated. Of all forms of nanotechnology, this is the most well established. "Bottom-up", or molecular nanotechnology (MNT), applies to building organic and inorganic structures atom-by-atom, or molecule-by-molecule.

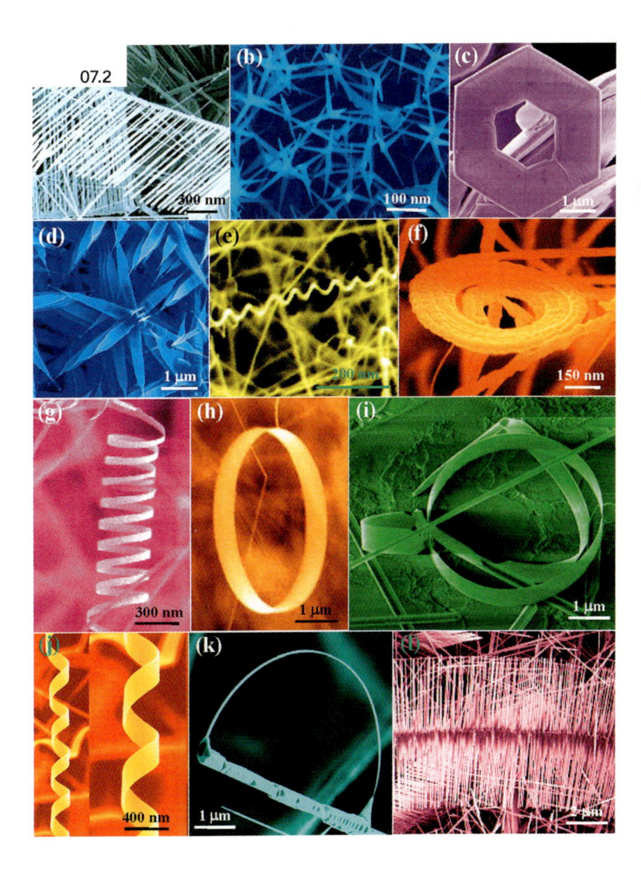

07.1 / GROWTH OF ALIGNED ARRAYS OF NANOWIRES IS OF GREAT IMPORTANCE IN NANOTECHNOLOGY.

07.2 / NANOSTRUCTURES COME IN A VARIETY OF UNIQUE SHAPES.

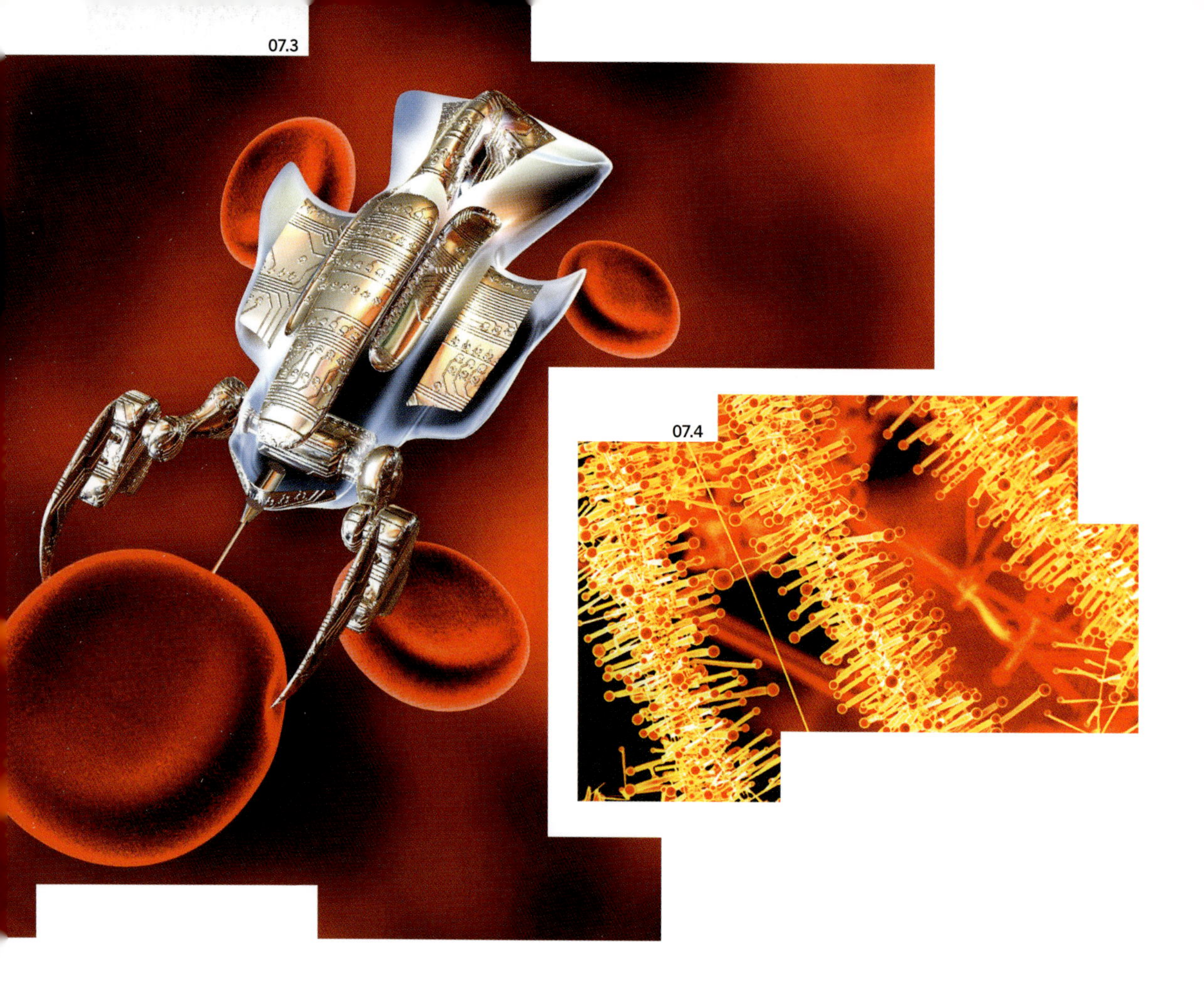

The following are the long-term goals of the NIH Nanomedicine Roadmap Initiative that are anticipated to yield medical benefits as early as 2020:

- Doctors have the ability to cure the initial cancer cells before they become tumors.
- Broken part of cells may be replaced with a miniature biological machine.
- Pumps the size of molecules implanted to provide life-saving medicine when necessary.

NIH will begin its effort by establishing Nanomedicine Development Centers, which will serve as the intellectual and technological centerpiece of the NIH Nanomedicine Roadmap Initiative. These centers will be staffed by highly multidisciplinary scientific teams including biologists, physicians, mathematicians, engineers, and computer scientists. Research conducted over the first few years will be directed toward gathering extensive information about the physical properties of intracellular structures that will inform scientists about how biology's molecular machines are built.

To help meet the goal of eliminating death and suffering from cancer by 2015, the National Cancer Institute (NCI) is engaged in efforts to harness the power of nanotechnology to radically change the way we diagnose, image, and treat cancer. Already, NCI programs have supported research on novel nanodevices capable of one or more clinically important functions, including detecting cancer at its earliest stages, pinpointing its location within the body, delivering anticancer drugs specifically to malignant cells, and determining if these drugs are killing malignant cells.

07.5

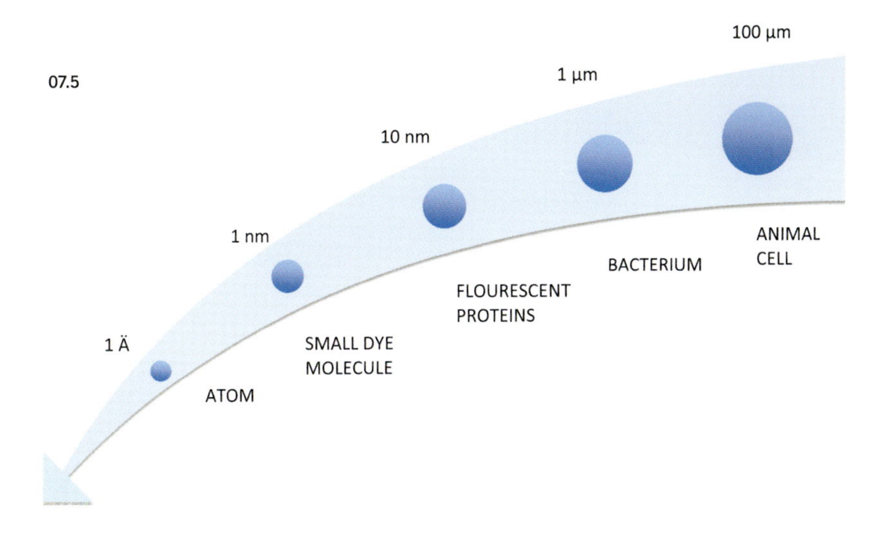

As these nanodevices are evaluated in clinical trials, researchers envision that nanotechnology will serve as multifunctional tools that will not only be used with any number of diagnostic and therapeutic agents, but will change the very foundations of cancer diagnosis, treatment, and prevention.

To successfully detect cancer at its earliest stages, scientists must be able to detect molecular changes even when they occur only in a small percentage of cells. This means the necessary tools must be extremely sensitive. The potential for nanostructures to enter and analyze single cells suggests they could meet this need.

Many nanotechnology tools will make it possible for clinicians to run tests without physically altering the cells or tissue they take from a patient. This is important because the samples clinicians use to screen for cancer are often in limited supply. Scientists would like to perform tests without altering cells, so they can be used again if further tests are needed. Reductions in the size of tools means that many tests can be run on a single small device. This will make screening faster and more cost-efficient.

Scientists are creating edible capsules only nanometers or billionths of a meter in size to enhance food or medicine. Edible nanoparticles are composed of materials either relatively inert in the body, such as silicon or ceramics, or materials that react with what bodies eat or with chemistry, such as polymers. One key advantage that edible nanoparticles have over larger particles is that breaking up something to the nano-level that is extremely insoluble in nature, like some drugs, can make it easier for them to be released in the body.

As just one example of how nanotechnology is helping to advance healthcare, a camera encapsulated in a large pill can visualize the gastrointestinal tract and eliminate the need for invasive diagnostic procedures. This capability points the way to next-generation nanotechnology devices that will uncover much greater information about the human body and allow medicine to be conducted with unprecedented precision.

Another example is Robert Langer's work at MIT where his team is beginning clinical trials of **"magic bullets" that employ nanotechnology to deliver drugs where necessary.** The nanoparticles are bio-degradable allowing organic molecules to bind to the desired target.

07.3 / NANOBOTS WITH RED BLOOD CELLS.

07.4 / NANOSTRUCTURES HAVE THE POTENTIAL OF MEASURING NANOSCALE FLUID PRESSURE AND FLOW RATE, SUCH AS BLOOD PRESSURE AND FLOW RATES.

07.5 / NANOSCALE COMPARISONS.

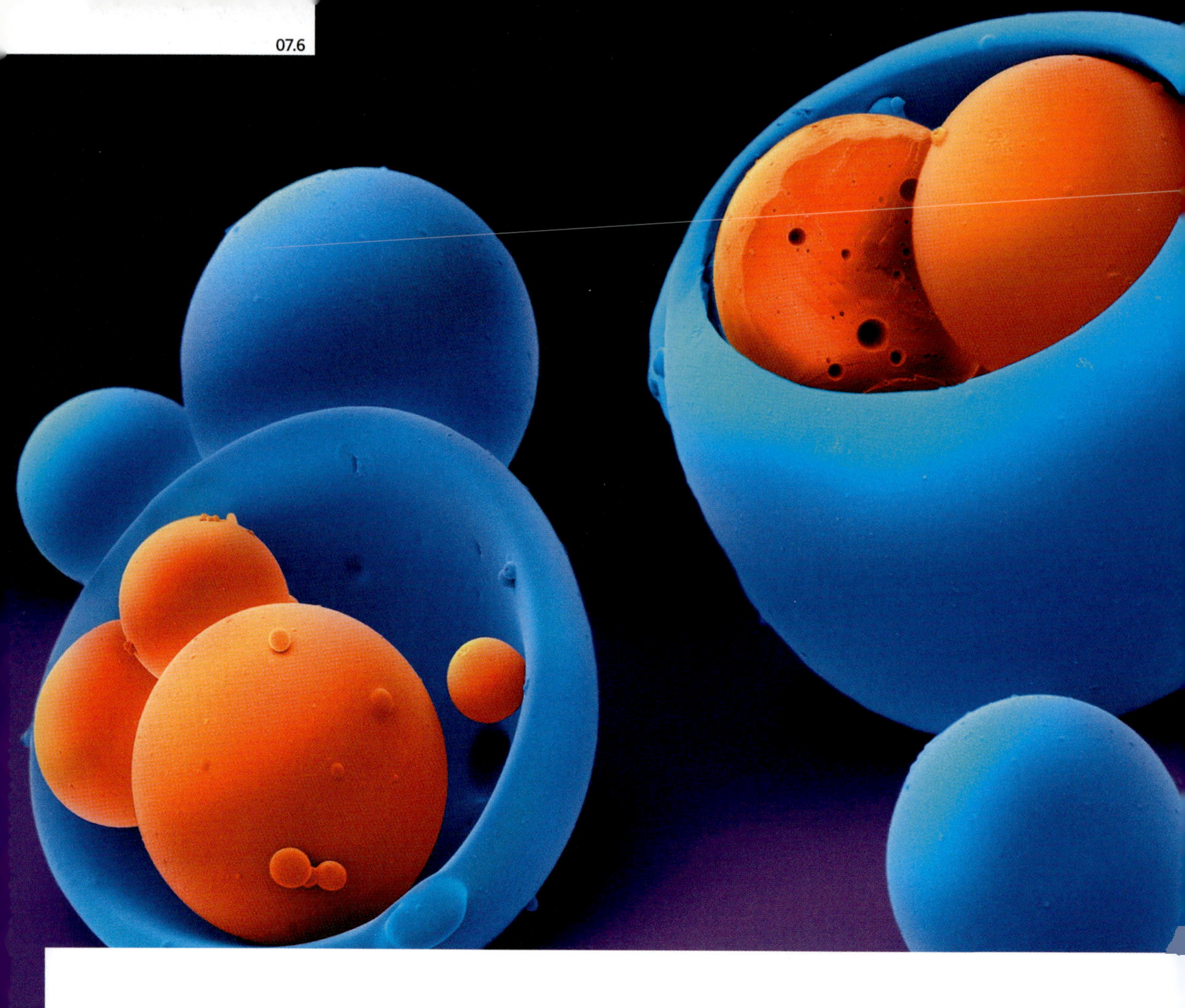

"In the future, nanoscale devices will run hundreds of tests simultaneously on tiny samples of a given substance. These devices will allow extensive tests to be conducted on nearly invisible samples of blood," said Langer.[4]

Opportunities for Other Research Discoveries /

"Nanoparticles, particles with dimensions on the order of billionths of a meter, could potentially be used for more efficient energy generation and data storage, as well as improved methods for diagnosing and treating disease. Learning how to control and tailor the assembly of these miniscule particles into larger functional systems remains a major challenge for scientists," say researchers at the Brookhaven National Laboratory.[5]

One of the most promising discoveries in the field of nanotechnology has been that of carbon nanotubes, which simply put are extremely thin (the diameter of a carbon nanotube is approximately 10,000 times smaller than a human hair) hollow cylinders made of carbon atoms. Carbon nanotubes possess unique properties; they are strong and resilient yet lightweight, with good thermal conductivity. These unique properties have enabled the possible use of carbon nanotubes in nanoelectronic and nanomechanical devices, among several other applications.

Hong Jie Dai's team at Stanford University has developed carbon nanotubes that can bring proteins and DNA into cells, which could potentially deliver drugs. "Nanodyne makes a tungsten-carbide-cobalt composite powder (grain size less than 15 nanometers) that is used to make a sintered alloy as hard as diamond, which in turn is used to make cutting tools, drill bits, armor plate, and jet engine parts," reported Nanotechnology Now.[6] Wilson Double Core tennis balls have a nanocomposite coating that keeps them bouncing twice as long as an old-style ball.

In recent years new breeds of microscopes made it possible to move atoms. Scanning Probe Microscopes can separate atoms by pushing them around on a surface. Scanning Tunneling Microscopes and Atomic Force Microscopes can

07.7

make surfaces with the dimensions of atoms visible. Soon instruments will be able to pick atoms and molecules up and move them. With this nanotechnology, the ability to control the arrangement of atoms is near. To move ahead, this technology revolution needs miniature-programmed assemblers (nanorobots). These devices would have a submicroscopic arm that would pop atoms into place with precision. This would enable them to build anything, including themselves. This type of manufacturing would give thorough and inexpensive control of the structure of matter.

If you make a plastic with nanotechnology, you can use feedstocks of pure elements such as carbon, hydrogen, and oxygen and force individual atoms deliberately into chemical bonds without taking the immediate steps. You could also build the plastic into the final shape you desired without injection molding. All "reactant" becomes "product" with no wasteful by-products.

Concerns /

Since 2000, awareness of nanotechnology among environmental activists, regulators, and lawmakers has been on the rise. Environmental organizations have expressed fears about the potential ecological and health consequences of mainstream nanotechnology, and have called for increased research into safety of nanoparticles.

History /

In 1959, physicist Richard Feynman suggested that it should be possible to build machines small enough to manufacture objects with atomic precision. Among other things, he predicted that information could be stored with amazing density. In the late 1970s, Eric Drexler began to invent what would become molecular manufacturing. He quickly realized that molecular machines could control the chemical manufacture of complex products, including additional manufacturing systems, which would be a very powerful technology. Drexler published scientific papers beginning in 1981. Although Norio Taniguchi coined the phrase "nano-technology" in 1974 to describe precision micromachining, it was Drexler who introduced the term "nanotechnology" to a wider audience in his 1986 book *Engines of Creation* to describe this approach to manufacturing and some of its consequences.

07.6 / MICROPARTICLE DRUG DELIVERY—EACH MICROPARTICLE IS APPROXIMATELY 20 MICROMETERS AND RELEASES A DRUG SLOWLY INTO THE BODY THROUGHOUT THE DAY.

07.7 / NANOPARTICLES USED FOR DRUG DELIVERY.

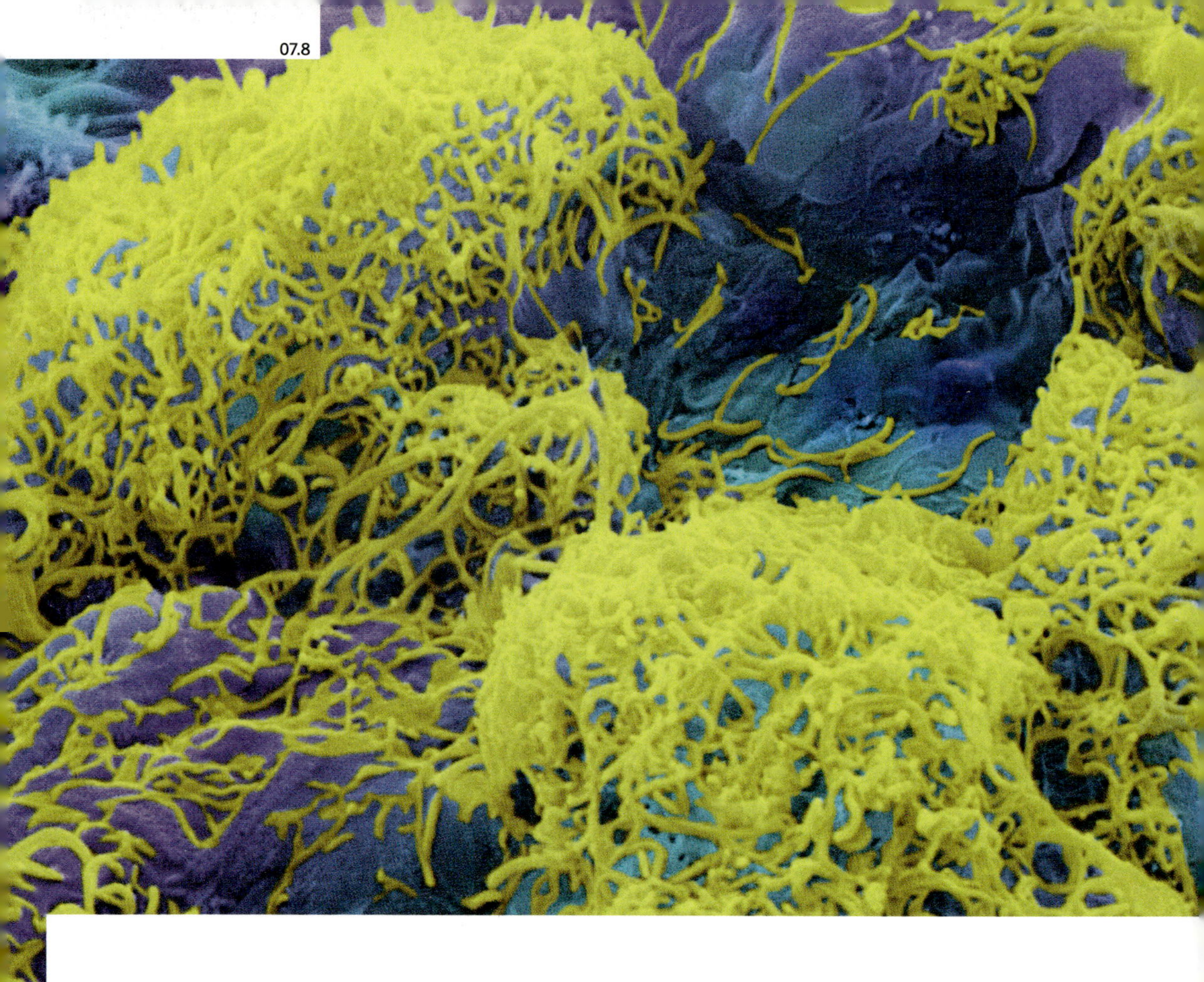

In 1985, researchers reported the discovery of the "buckyball," a lovely round molecule consisting of 60 carbon atoms. This led in turn to the 1991 discovery in Japan by Sumio Iijima of a related molecular shape known as the "carbon nanotube"; these nanotubes are about 100 times stronger than steel but just a sixth of the weight, and they have unusual heat and conductivity characteristics that guarantee they will be important to high technology in the coming years. Since this landmark discovery by Iijima, a leading researcher in the field of nanotechnology in research laboratories at Japanese corporation NEC, Japan has emerged as one of the leaders in nanotechnology research and development.

In 1992 Drexler published *Nanosystems*, a technical work outlining a way to manufacture extremely high-performance machines out of molecular carbon lattice ("diamondoid"). Drexler established the field of molecular nanotechnology. His work showed quantum chemists and synthetic chemists how the knowledge of bonds and molecules could serve as the basis for further development of manufacturing systems of nanotechnology, and showed physicists and engineers how to scale down their concepts of macroscopic systems to the level of molecules.

In 2004 Douglas Connect announced the launch of InterNanotech, a new online community for international researchers in computational and experimental nanoscience and nanotechnology. **By holding regular seminar sessions over the Internet, supported by virtual communication and networking tools, InterNanotech is designed to facilitate rapid communication of new research results and discoveries between different disciplines and geographic regions.**

International Research and Development /

In 2003 just 14 countries accounted for 90 percent of the $5.5 billion invested in nanotechnology. The USA and Japan provided over $3 billion of that investment alone. NanoChina was launched in early February 2006 to provide a wide range of market-specific nanotechnology information services for businesses, researchers, and the public sector in China. NanoChina will act as a bridge between the nanotechnology activities that are

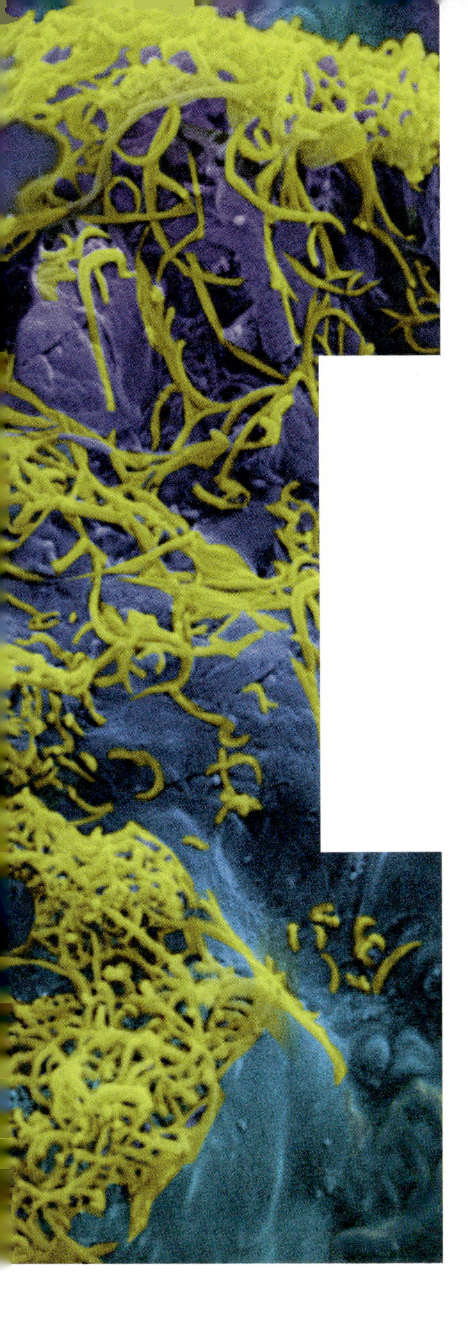

taking place in China and the rest of the world, with the aim of disseminating information and setting up nano-business and networking opportunities.

In the Chinese government's National Long Term Development Layout, 2006–2020, nanoscience and nanotechnology is one of the four national research programs. Looking forward, the research and development of nanotechnology will continue to be one of the most active fields in China.

The nanotechnology research in China covers more aspects of nanoscience and nanotechnology. Four areas have generally been addressed:

- First, synthesis and growth of nanoscale structures composed of semiconductors, metals, and polymers, and to make electronic devices and sensors from molecules and nanoparticles.
- Second, characterization of nanostructures and imaging electrons inside nanostructures by various analysis methods, especially scanning probe microscopes (SPMs).
- Third, manipulation and control of the quantum state such as spins, charges, and orbit of the molecule and nanostructures.
- Fourth, the application of nanomaterials and nanotechnology in biology and life science, such as drug carriers and therapy.

The researchers have so far gained many accomplishments mainly on the assembly, characterization, and manipulation of nanostructure via "bottom up" stratagem. However, the design and the fabrication of nanodevices have not been adequate due to the incomplete "top down" techniques based on the weak micro-fabrication process.

07.8 / CARBON NANOTUBES, 150 NANOMETERS IN DIAMETER, ARE PART OF A DRUG-DELIVERY SYSTEM.

Japan /

In addition to government-sponsored research and development in the field, large Japanese corporations and universities have dedicated millions of dollars to nanotechnology research. Hitachi, Sony, Toray, Mitsubishi, Fujitsu, and Mitsui are among the Japanese corporations with nanotechnology initiatives. Mitsubishi plans to come out with nano-enhanced, energy-saving, flat-panel technology for computer displays and televisions. At Fujitsu, research is being conducted to build circuits using nanotubes.

Japan views the development of nanotechnology as the key to restoring its economy, according to a report by the *Journal of Japanese Trade & Industry*. But the country faces stiff competition from the US and Europe, both of which have increased their research-and-development budget for nanotechnology. Government funding for nanotechnology in the US increased 9.5 percent to $847 million, up from $774 million in 2003. This figure is less than 2 percent of what is spent on biodefense. Following closely is the UK, which recently announced $150 million in funding for nanotechnology research over the next six years. Japan, however, has thrived in the nanomaterials and nanoscale fabrication sectors, driving products to the market at a faster rate than the US.

"Russia will pour over 1 billion US dollars in the next three years into equipment for nanotechnology research as it uses massive oil and gas export earnings to seek to diversify an economy now heavily dependent on raw materials," First Deputy Prime Minister Sergei Ivanov said in 2008.[7] **"Nanotechnology is a very promising scientific and technical field, capable of fundamentally changing the model of the Russian economy ... from a fuel economy to an economy of the future,"** Ivanov also stated. Russian Prime Minister Vladimir Putin has repeatedly stressed the importance of freeing the economy from its dependence on commodities markets and turning it toward a more high-tech, knowledge-based model.

Funding /

"The private sector is investing in nanotechnology R&D at a level comparable to or greater than the Federal R&D investment. Large companies in industries ranging from aerospace and automotive to chemicals and semiconductors are moving to incorporate nanotechnology into their products and processes. The venture capital community is making significant investments in nanotechnology-based start-ups, leading to a growing number of 'pure play' nanotechnology businesses, many of which are developing entirely new products and approaches, for example, in medical imaging and therapeutics."[8]

The National Nanotechnology Initiative (NNI) is a multi-agency US Government program aimed at accelerating the discovery, development, and deployment of nano-meter science, engineering and technology. The 2008 budget provides nearly $1.5 billion for the NNI, more than triple the estimated $464 million spent in 2001. The NNI budget provides continued support for basic research to understand nanoscale phenomena.

International collaborations in non-competitive and pre-competitive areas of nanotechnology continue at an accelerated pace. In 2006, the Organization for Economic Cooperation and Development (OECD) established a Working Party on Manufactured Nanomaterials, chaired by the USA, to address health and safety issues. With US leadership, a second OECD working party is being formed under the Committee for Scientific and Technological Policy to address broader issues related to realization of the benefits of nanotechnology, including assessing economic impact, education and training, and public communication. The National Technology Coordination Office (NNCO) and NNI agencies are also collaborating with the nanotechnology research institute in Belgium to organize a workshop on public outreach and communication.

07.9

The NNI continues to develop state-of-the-art facilities and infrastructure. In 2008, the fifth of five DOE Nanoscale Science Research Centers, the Center for Functional Nanomaterials at Brookhaven National Laboratory, will become fully operational. These centers provide the broad research community access to facilities, instrumentation, and expertise in support of advanced nanotechnology research and development. With NSF funding, the center for Learning and Teaching at the Nanoscale and the Nanoscale Informal Science education Network both unveiled "one-click resource" websites complementing their growing collaborative networks, while the traveling exhibit "Too Small to See" was visited by over 350,000 children and their families in its first 90 days at the Walt Disney World's Epcot Center.

In *NanoFrontiers: Visions for the Future of Nanotechnology*, Karen F. Schmidt summarized the opportunities of nanotechnology: "Perhaps what now seems almost like science fiction will one day seem like a historic paradigm shift that helped us solve some of our most pressing and complex problems ... it is impossible to imagine what will emerge as researchers, engineers, and social scientists continue to interact with one another. Nanotechnology will almost certainly evolve in ways that we cannot predict, it will change our world and it will become a key part of the advance of human potential."[9]

07.9 / ARTIST'S CONCEPTION OF NANOPARTICLES.

07.10

07.11

I had the opportunity to meet with Dr. Zhong Lin Wang, the director for the Center for Nanostructure Characterization at Georgia Tech, for an afternoon to get his views on nanotechnology. Dr. Wang emphasized four key areas of nanotechnology research that he believes will take place over the next 10 years.

The first area of research focuses on utilizing energy efficiently. For example, only 2 percent of the energy that comes from the coal used to generate electricity in a power plant produces light from a bulb, the other 98 percent of the energy is lost. Nanotechnology can help to provide better ways of transmitting energy, reducing heat loss, and allowing a higher percentage of the energy to be utilized.

Renewable and green energy technologies are being developed as sustainable energy solutions and these too will benefit from developments in nanotechnolgy. There are also research opportunities to power small electronic devices. For instance, Dr. Wang showed me nanogenerators that convert mechanical energy from physical motion into electricity. One day, with the benefit of a nanogenerator, the act of walking will provide enough power to charge a cell phone.

The second area of research focuses on biomedical science. The early detection of infection and disease can be identified by using nanodevices and nanorobots inside the body. Single-cell diagnostic technology will become a key part of a proactive healthcare system, possibly within the next 10 years. Imaging has grown significantly within the past 10 years. The equipment we have today can record three-dimensional images all the way down to the nanoscale of an individual cell.

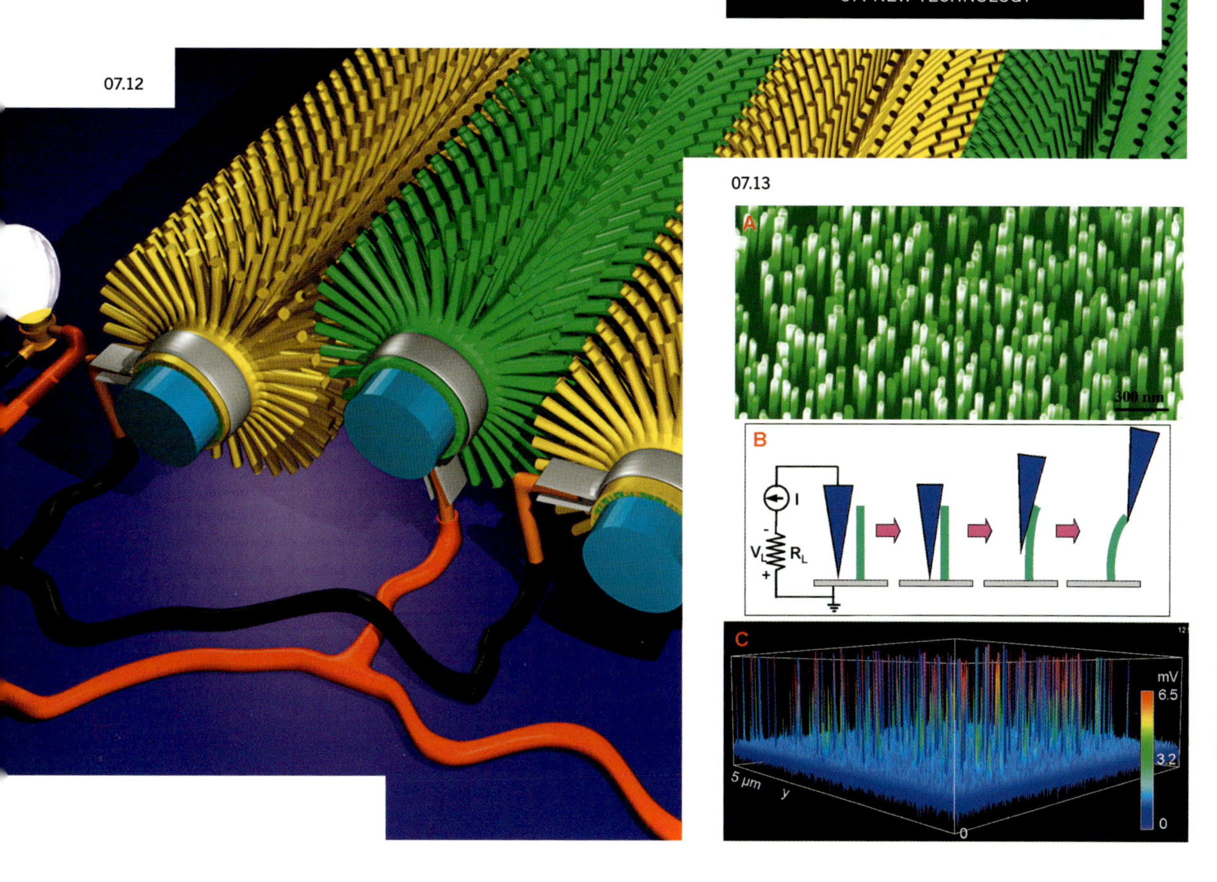

07.10 / NANOBELTS OF ZINC OXIDE DEFINE A NEW GROUP OF NANOMATERIALS FOR APPLICATIONS IN NANOELECTRONICS, NANOSENSORS, NANOACTUATORS, AND BIOSENSORS.

07.11 / ALIGNED PROPELLER ARRAYS OF ZINC OXIDE, WHICH HAVE APPLICATIONS IN NANOSCALE SENSORS AND TRANSDUCERS.

07.12 / A SCHEMATIC ILLUSTRATION OF THE MICRO-FIBER-NANOWIRE HYBRID GENERATOR, WHICH IS THE BASIS OF USING FABRICS FOR GENERATING ELECTRICITY.

07.13 / SCHEMATIC EXPERIMENTAL PROCEDURE FOR GENERATING ELECTRICITY FROM A NANOWIRE USING A CONDUCTIVE ATOMIC-FORCE MICROSCOPE.

Nanotechnology advances within the biomedical sciences will take longer to reach the market than those in other areas of nanotechnology research because of the approval process required before medical products can be marketed. In the short term, there should be a better financial return on discoveries utilizing more efficient energy use because the approval process will take less time.

The third area of research focuses on environmental monitoring. More efficient and cost-effective research tools are being developed to measure acid rain, air, and water pollution, as well as the migration of animals. For instance, sensors as small as a thumbnail can be placed on animals to track where they are traveling, or nanotools will be able to track the quality of a city's water supply for quick response in case of a terrorist attack. **The fourth area of research focuses on designing smaller, more efficient, and less costly electronics from computers to MP3 players.**

Dr. Wang explained that approximately 10 years ago, just a handful of faculty members at Georgia Tech were involved in researching nanotechnology. Five years later, a survey indicated approximately 450 of the 1,000 faculty members were conducting some research related to nanotechnology. For the long term, goals and projects in nanotechnology need to be more crystallized and hopefully the research delivers on those projects and products. The research should aim to solve significant problems to help humankind. A change occurred recently, driven by nanotechnology research, which requires multi-disciplinary teams; global research and collaboration. Multiple areas of research expertise need to work together. The education system is now changing to support new research models. **Students will need to have an area of expertise, know more disciplines, and creatively work as a team—both locally and globally.**

07.14

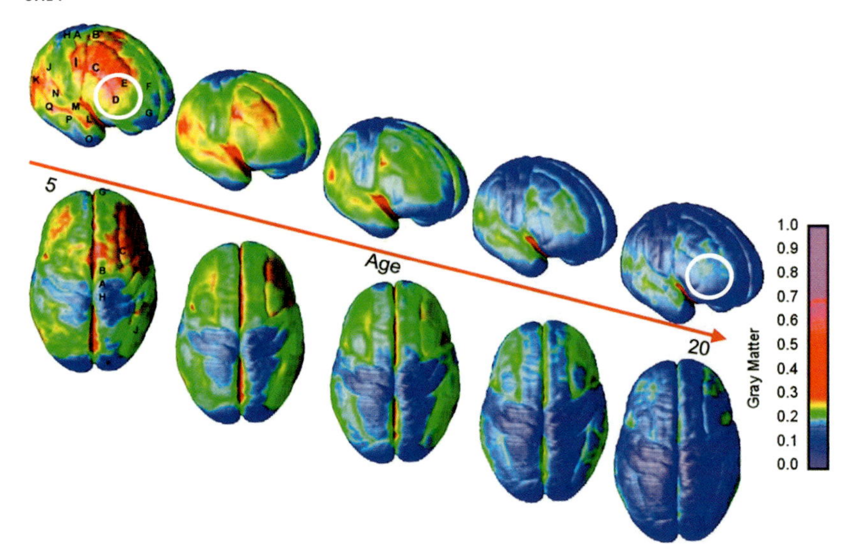

Neuroscience /

Science /

Neuroscience is the study of nervous systems and of the nerve cells that make up nervous systems. The human brain contains one-hundred-thousand-million such cells. How the brain works remains the most alluring and baffling of all questions on the frontier of understanding. Can the brain understand the brain? Can it understand the mind? Is the brain some kind of giant computer or something more?

Only in recent decades has neuroscience become a recognized discipline. It is now a unified field that integrates biology, chemistry, and physics with studies of structure, physiology, and behavior, including human emotional and cognitive functions. Neuroscience research includes genes and other molecules that are the basis for the nervous system, individual neurons, and ensembles of neurons that make up systems and behavior.

It is increasingly important that neuroscientists construct theories of nervous function that encompass all levels of organization, from molecular events to behavior. Also, malfunctions of nerve cells lead to baffling conditions ranging from schizophrenia and senile dementia to diseases of motor control like Parkinson's disease, and to changes in mood that manifest as mania or depression. Many of these conditions are treated with drugs, hence pharmacology is a further essential component of neuroscience. Some conditions that result from neuronal degeneration or spinal injury may be helped by neuron transplantation. This very modern concept of treatment is in its infancy.

The human body contains roughly 100 billion neurons, which are the functional units of the nervous system. Neurons communicate with each other by sending electrical signals and then releasing chemicals called neurotransmitters that cross synapses—small gaps between neurons.

In attempting to understand the brain, neuroscientists are constantly trying to link understanding at different levels of analysis. To understand the behavior of organisms it is necessary to understand the behavior of neurons. Consequently, neuroscience employs

07.15

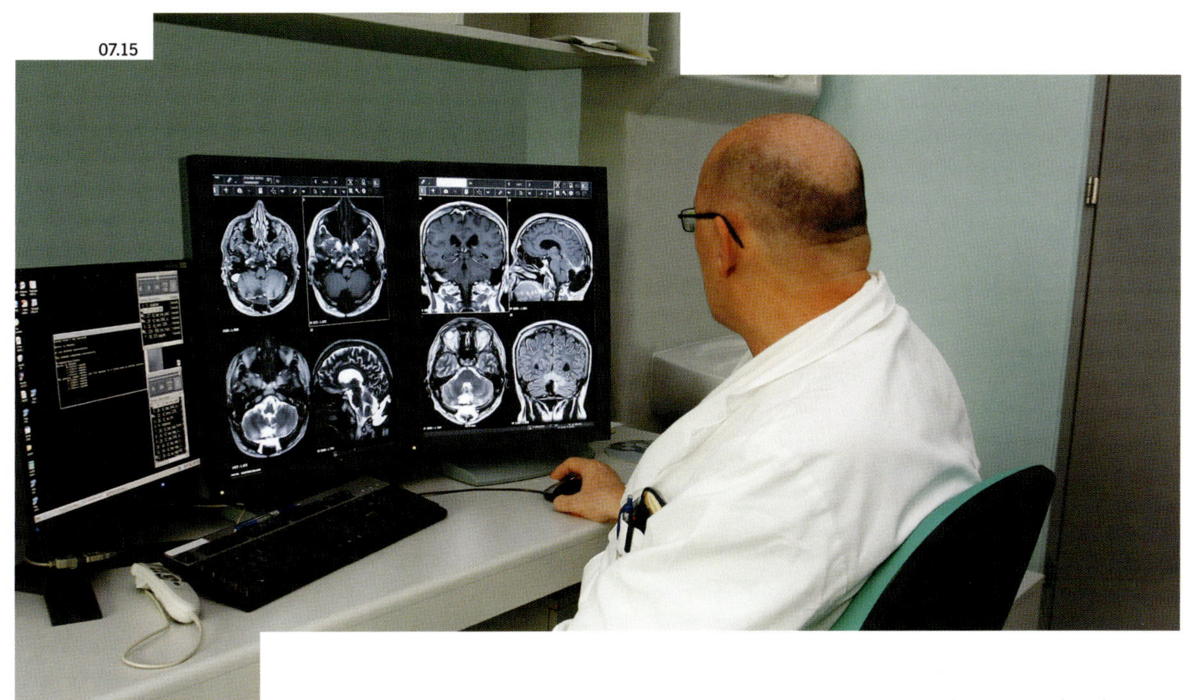

a multidisciplinary approach. Eventually, neuroscience research will have a biological explanation for even the most complex behaviors.

Neuroscience research can include the study of cellular and molecular neurobiology, neuronal survival and death, development and regeneration of brain cells, cell to cell communication, and the study of receptors and their modulation. Through these investigations, neuroscience research aims to understand diseases such as stroke (deprivation of parts of the brain from its blood supply), difficulty with memory function (Alzheimer's disease), abnormal-movement disorders (Parkinson's and Huntington's diseases), and disorders of white-matter function (multiple sclerosis). Neuroscience research also includes work on optic nerves, cardiac function, sense of smell, difficulty in swallowing, weakness in the muscles, and other neural-based diseases and health problems.

Neuroscientists are utilizing computer technology to expedite the move toward a better understanding of the complex inner workings of the brain. It is the advances in this computer technology that are fundamentally changing the discovery process with still unknown, but certain beneficial outcomes to medicine, and ultimately to human health.

07.14 / BRAIN DEVELOPMENT THROUGH EARLY ADULTHOOD, WITH BLUE INDICATING THE MATURE STATE. THE PREFRONTAL CORTEX (WHITE CIRCLES), WHICH GOVERNS JUDGMENT AND DECISION-MAKING FUNCTIONS, IS THE LAST PART OF THE BRAIN TO DEVELOP.

07.15 / RESEARCHER REVIEWING BRAIN IMAGES.

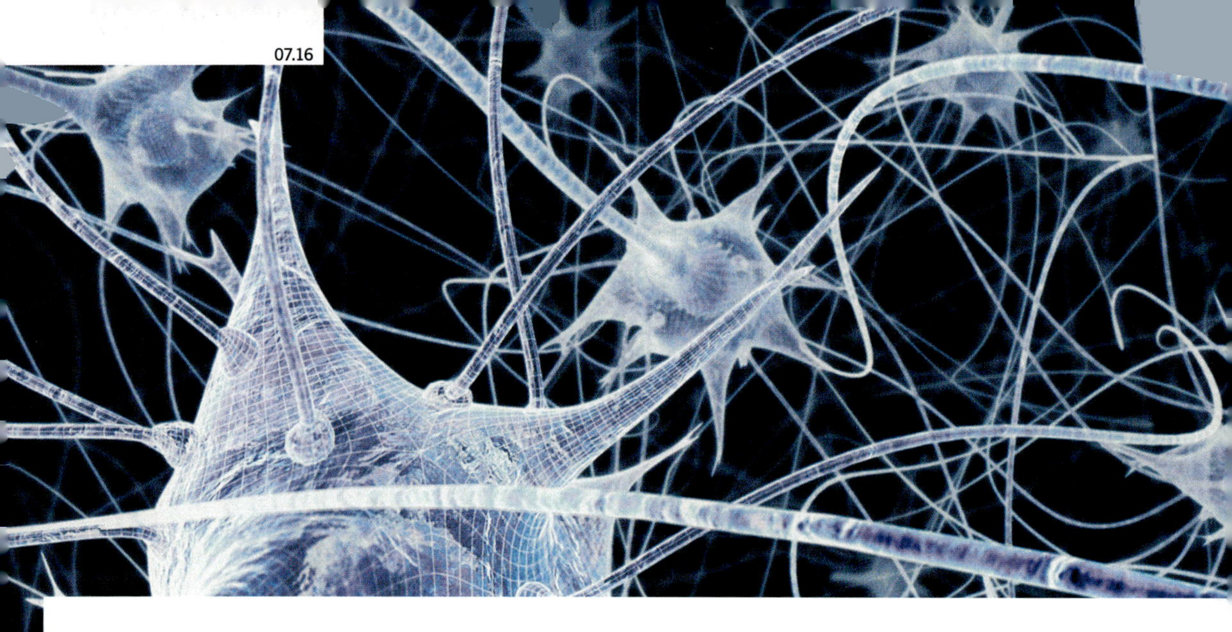

There are many areas of study within neuroscience including:

- Neurobiology, which aims to study cells of the nervous system and the organization of these cells into functional circuits that process information and mediate behavior.
- Molecular and cellular neuroscience, which integrate neurobiology with neurochemistry with the goal of understanding the cellular and chemical mechanisms of normal and abnormal brain function.
- Cognitive neuroscience, which aims to understand the mechanisms that underlie "higher level" brain functions, usually in humans. These include language, learning and memory, attention, and emotion.
- Computational and systems neuroscience, which seeks to understand how information is processed by the nervous system. The methods of research combine mathematical and computational models with physiological recordings of single cells, neuronal clusters, and entire brain systems. "The rapidly growing field of computational neuroscience holds great promise for enhancing the understanding of the functions of genes and proteins within nerve cells, as well as the interactions responsible for storing information and generating behavior. Pairing computer science with biomedical research facilitates deeper insight into mechanisms of complex biological systems, including the most complex—the brain," reported the Neuroscience Research Center at the University of Texas Health Science Center at Houston.[10]

Opportunities for Cures /

Researchers are dedicated to understanding the underlying causes of a wide variety of neurological disorders in order to develop new treatments and cures, including:

- Alzheimer's disease and dementias
- Cerebellar ataxia
- Cerebral palsy
- Dizziness, vertigo, and balance disorders
- Epilepsy and tremor
- Gait and walking disorders
- Head injury
- Huntington's disease
- Hypertension
- Learning disorders
- Memory loss
- Movement disorders
- Multiple sclerosis
- Neuronal (brain cell) dysfunction
- Pain
- Parkinson's disease
- Stroke
- Synaptic dysfunction
- Vestibular disorders—Meniere's disease
- Eye disorders

More than 600 disorders afflict the nervous system. Neurological disorders impact an estimated 50 million Americans each year, creating an annual economic cost of hundreds of billions of dollars in medical expenses and lost productivity. Brain and spinal cord injury caused by traumatic accident is the leading cause of disability and death in children and young adults. In the US alone, one head injury occurs every 15 seconds and one death and one permanent disability from traumatic brain injury every five minutes. "Neurological disorders affect millions of people around the world. In Canada alone, neural injury and disease strike more than four million people and cost $30 billion annually in healthcare costs and lost productivity. For most of these disorders, treatment is still either unsatisfactory or non-existent," wrote K. Chaundy in *Why is Neuroscience Important?*[11]

07.17

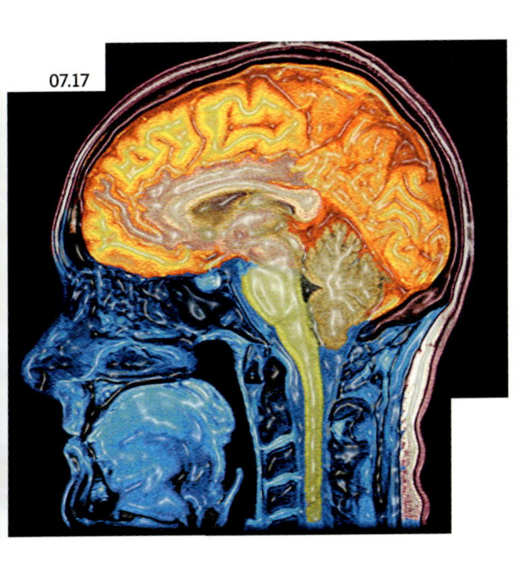

Blue Sky Research /

In 2009, the National Institute of Neurological Disorders and Stroke (NINDS) invited a distinguished panel of researchers, clinicians, and lay members to develop a "Blue Sky Vision for the Future of NeuroScience." The following broad topics were identified:

- Enable early and routine diagnosis of neurological conditions
- Develop new therapeutic strategies
- Accelerate the process of therapy development
- Understand the healthy nervous system
- Understand neurological disease
- Develop new technologies for observing the nervous system
- Develop new strategies to probe neural functions
- Improve strategies for data collection and analysis.

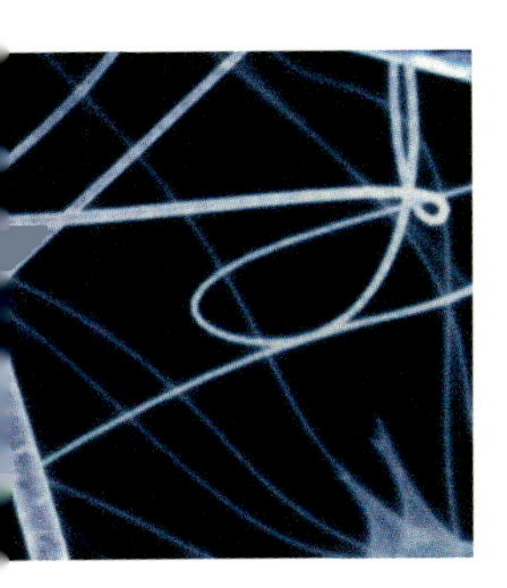

Fifteen years from now, the hope is to have better interventions for all neurological conditions, and to accomplish this NINDS will systematically review basic, translational, and clinical opportunities. I had the opportunity to talk with Dr. Robert Finkelstein at the NINDS. He shared with me his thoughts, many of which are being addressed by NINDS and other NIH institutes.

Dr. Finkelstein says that **as the baby-boom population ages, there will be a huge increase in the incidence of Alzheimer's disease and other neurodegenerative disorders.** It is important that NINDS and other components of the NIH address this challenge, as well as many other issues, by using basic and applied research. To do so, it is critical that NIH ensure that it continues to maintain the research pipeline, and by continuing to attract promising new investigators into neuroscience and other areas of biomedical research. As we learn more about the molecular and genetic causes of neurological disease, personalized medicine will become increasingly important. Another key challenge will be storing and using the tremendous amount of data that is being generated through DNA sequencing, imaging, and other diagnostic techniques.

Another important goal for neuroscience will be identifying the totality of connections in the human brain. The Human Connectome Project is one of three Blueprint Grand Challenges currently being planned through the NIH Blueprint for Neuroscience Research, a consortium that includes NINDS and the other NIH institutes that fund neuroscience research. Like the Human Genome Project, this is an ambitious endeavor that will help us to understand both normal and diseased brain function. A second Grand Challenge involves understanding chronic pain, and a third focuses on attempting to develop effective treatments for specific diseases of the nervous system.

As the field progresses, it will become increasingly important to study neuroscientific phenomena not just in isolation, but as part of a larger whole. Molecular and cellular data must be integrated so that we can gain a better understanding of higher brain functions. This systems approach will ultimately allow researchers to develop a more comprehensive understanding of how the brain works. In the 21st century, this understanding will form the basis for developing therapies for treating the many devastating diseases that affect the brain.

07.16 / NEURONS IN THE BRAIN.

07.17 / ENHANCED MRI SCAN OF A HEAD.

08. IMPROVING THE SCIENCE OF SCIENCE

Introduction /

In recent years we have re-invented our approach to research. Because teamwork in and between laboratories is now emphasized, we have created infrastructures to support human interaction. Furthermore, equipment has become much more efficient, speeding advances in fields such as virtual reality. These developments are being driven by competition and the urgent need to find solutions to pressing world problems.

Interdisciplinary Research /

The key to successful interdisciplinary research is collaboration.

But collaboration is much more than people from various backgrounds sharing ideas in informal settings. Mass collaboration on the Internet is greatly accelerating ingenuity and intelligence sharing.

Of late, many institutions from undergraduate programs to leading federal laboratories, have shifted from the departmental model to the team model. Cross-pollination among disciplines works. Together cell biologists, engineers, and materials scientists have made it possible to grow complex human tissues that can repair the human body. Interdisciplinary research, supported by three of the seven Research Councils UK, has produced tissue scaffolds that can be injected. These tissue scaffolds are now being commercially developed by Critical Pharmaceuticals Ltd., a Research Councils UK spin-off company that recently secured £1 million of venture-capital investment.

Information technology and computational sciences are revolutionizing other disciplines, from environmental sciences (including climate-change modeling) to bioinformatics and systems biology (using advanced computational techniques to analyze complex biological processes) to medicine. Magnetic resonance imaging, which provides doctors with images of living tissue, emerged from

08.2

interdisciplinary research linking physics, medicine, and engineering at the University of Nottingham. Nearly all hospitals now have MRI scanners, which are revolutionizing medical diagnosis and brain science.

This chapter presents case studies of new world-class laboratory facilities, of which many more are needed. Designs for such laboratories are changing rapidly.

The Bio-X Program /

Project: James H. Clark Center, Stanford University, Stanford, California

Gross square feet: 245,000

Construction cost: $100 million

Year completed: 2003

By its very design, the Clark Center continually builds new alliances in science, notes Beth Kane, director of Bio-X Operations. Work areas are flexible, inviting, and provocative. Open spaces, rather than cubicles, are filled with energy and activity. Generous space and equipment are dedicated to collaborative work. Each floor features conference and seminar rooms, and a restaurant and coffee shop beckon interaction. At its core, the building design recognizes that gatherings in social settings are profoundly important science incubators.

The Stanford University Bio-X program facilitates multidisciplinary research connected to biology and medicine. "Bio" stands for today's dominant science, biology, which explores everything from an enhanced understanding of the mechanisms of the human brain, to how a cell functions, to the human genome. The variable "X" stands for engineering, chemistry, physics, computation, medicine, and other sciences. Taken together, Bio-X fosters "the coming together of leading-edge research in basic, applied, and clinical sciences to enable tomorrow's discoveries and technological advances across the full spectrum, from molecules to organisms," the program states.[1] It is through innovation, collaboration, and interaction that this goal will be achieved.

08.1 / THE CLARK CENTER, STANFORD UNIVERSITY, STANFORD, CALIFORNIA.

08.2 / CHALLENGES FACED BY LIFE SCIENCE RESEARCHERS. RESEARCH TAKES TIME, AND POSSIBLY THE BIGGEST CHALLENGE RESEARCHERS FACE IS CONVINCING INVESTORS THAT TIME EQUALS SUCCESS.

Vision, Innovation, Collaboration, and Interaction /

The Clark Center design introduced new ways of fostering interdisciplinary collaborations and aligning unexpected partnerships in innovative research. The building resulted from creative dialogue between the founding Bio-X executive committee, which wanted to change the way scientists interact.

Instead of housing traditional laboratories and an internal core with many divisions, the building plan was turned inside out. Open interconnecting labs are on the interior, and lab and building support functions are outward and in the long, straight wings of the buildings.

The large labs accommodate, side by side, scientists representing schools from medicine to science and engineering to humanities and beyond. Because this required an ultra-flexible laboratory layout, the labs feature adaptable benches with utilities fed from above. With a truly adaptable floor plate, the building accommodates the varied lab types of today and provides the flexibility to meet the requirements of tomorrow's not-yet-conceived research labs.

Social and educational spaces are mixed throughout the building. An emphasis on placing the labs near social spaces recognizes that much of scientific discovery and dialogue happens outside of the lab. The Clark Center's distinctive design makes collaboration unavoidable. With inner courtyards walled completely in glass, researchers on different floors can easily look across the courtyard into other laboratories and observe their colleagues at work. The wings of the building are connected by bridges, and a raised platform in the center of the courtyard can be used as a stage for lectures or concerts.

The building also engages the larger campus by placing the laboratories on display to the courtyard. This raises the public face of the work in the building and inspires students to find out what is going on inside. The courtyard creates a hub for the campus and expresses the Bio-X vision of being a program that is much larger than the science inside one building.

The three wings that house researchers from dozens of departments foster a clear sense of energy. Researchers navigate freely between wet labs, outfitted with fume hoods and safety stations, and dry biocomputational spaces. Meeting rooms for brainstorming are never far away. Lab benches are scattered throughout the transparent building to facilitate visiting researchers for up to six months.

Innovative workspaces include accessible utilities, mobile casework, and adaptable infrastructure. Flexibility is

08.4

evident in the sealed epoxy floors, which allow work areas to be utilized as wet or dry labs, and a Unistrut ceiling and racks that provide drop-down access to standard utility services, such as gas, air, vacuum, electricity, communication, and water. These building services come down from the ceiling through a stainless steel guide to 7 feet above the workbenches. By loosening a few bolts, the mobile Unistrut drops can easily be moved. Sinks are the only items that are not mobile.

All casework is on wheels, so equipment or experiments can be quickly and easily moved, and disconnect valves are readily available for all utilities. A customized T-shaped docking station meets the needs of the researchers with two levels of shelves. Storage shelves can also be hung from the ceiling. In addition, copper feeds and fiber lines are available to meet high-speed computation needs. A 4-foot building services area above the ceiling includes the ductwork, sprinklers, and smoke detectors.

08.5

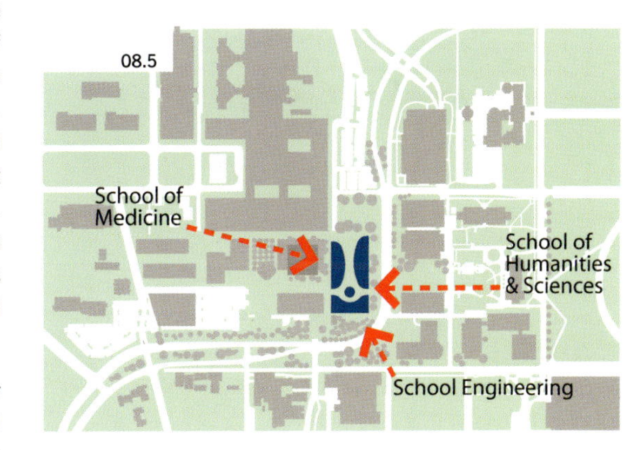

The laboratory is designed for change, which allows individual lab groups to rearrange their labs to suit their needs. Users can make small changes, such as reorienting a bench or moving a desk by a few feet. Larger changes, such as relocating a lab bench by more than a few feet or the complete reconfiguration of a lab, are organized through the building manager, which ensures that new layouts meet the safety requirements.

The laboratory benches and storage areas are based on a modular system that can be interchanged and moved around the building. Individual tables function as lab benches, allowing varied utility connections and storage options at the bench. Shelves are all based on the 3-foot module and are completely interchangeable.

08.3 / THE CLARK CENTER IS AN EXAMPLE OF A BUILDING PLAN TURNED INSIDE OUT.

08.4 / THE LARGE INTERNAL LABS HAVE A SERIES OF ADAPTABLE BENCHES WITH UTILITIES FED FROM ABOVE.

08.5 / THIS DIAGRAM SHOWS HOW THE CLARK CENTER'S THREE SCHOOLS ARE INTERCONNECTED AROUND THE CENTRAL COURTYARD.

08.6

Interaction areas /

Interaction is encouraged with regionally shared facilities, a restaurant, a café, seminar rooms, and an auditorium. Fifteen percent of the building is dedicated to public functions, such as conference rooms and classrooms, while 9 percent consists of facilities that are shared by people within the Clark Center. The restaurant is located on the first floor to encourage people to enter the building, and the seminar rooms are situated on the third floor to encourage visitors to explore the rest of the facility. A Peet's Coffee café is located on the third floor.

Called LINX, the family-style restaurant provides long, lab-coat white tables that are ideal for breaking down social barriers between people, Beth Kane, director of operations at Bio-X, tells the *Stanford Report*: "'At a long table, diners are more likely to strike up a conversation.' With its phosphorescence-green walls and long glass window opening up in the courtyard, the restaurant is nearly impossible to miss. 'That was extremely intentional,' Kane said. 'We wanted every facet of this building and this program to be welcoming and to expose people to new ways of thinking, eating, and doing.' "[2]

Alfred Spormann, associate professor of civil and environmental engineering, said that his best Clark Center collaborations so far have occurred outside the lab, for example along the balcony or inside Peet's Coffee. "If I go to a lab, I have a specific purpose," he said. "On the way, I run into people I usually don't see. The layout of the Clark Center is quite inviting for those interactions."[3]

Harvard Commits to Interdisciplinary Research /

As the lines blur between disciplines, such as biology and chemistry, universities must radically alter their approach to science. Harvard University is reinventing itself in hopes of maintaining a competitive edge. Determined to create a truly interdisciplinary science environment, Harvard secured an initial commitment of $50 million from the Harvard University Corporation in early 2007. The funding targets the university's declared need to make sweeping changes. **Harvard must "promote collaboration and respond rapidly to emerging research opportunities,"** the university's planning committee for science and engineering stated in July 2006.[4] Institutional agility is key.

The money will create new interdisciplinary departments, fund research, and pay for new laboratories and equipment for research that crosses traditional boundaries, reported the *Boston Globe*.[5] The university is positioning itself to respond quickly to new opportunities, including the following:

Using stem cells to decipher how organisms develop and to revolutionize medicine.

The convergence of biologists, physicists, chemists, and engineers dedicated to understanding the fundamental principles that explain the organization, reproduction, function, and evolution of biological systems, and using this knowledge to advance healthcare as well as engineering.

Using powerful arrays of computers to provide better links between large data sets and theories that seek to explain them.

New ways of understanding the details of the evolution and diversity of living things that would allow us to better protect our planet.

Combining basic science, engineering, and public policy to make and implement plans for sustainable energy generation and consumption.[6]

In the newly formed Department of Developmental and Regenerative Biology, research ranges from basic science to medical applications. This university-wide department is a historic first for Harvard, bringing together researchers from the medical school and the faculty of arts and sciences.[7] Other proposed cross-school departments include systems biology, chemical and physical biology, and neuroscience.

Improved technology infrastructure is another focus. Noting that dispersed geography and outdated IT equipment make it "difficult to know what other research is taking place across the university or what lab and research capabilities exist in other departments, schools, or hospitals," **the committee called for university-wide databases and video conferencing capability to inform its researchers about other projects on campus.**

The changes also emphasize an expanded science curriculum and more hands-on learning for undergraduates. "The last decade has seen an explosion of new life-sciences knowledge and methods, which has led to new fields such as genomics, systems biology, neuroscience, and evolutionary-developmental biology," Douglas A. Melton, Harvard natural sciences professor and life science council chair, told the committee. "These innovations are having profound impacts on fields spanning the breadth of life sciences, from the social sciences to the physical sciences. While Harvard faculty are actively engaged in research in these new directions, our current curriculum is not appropriately designed to permit undergraduates to learn about and get involved in these new fields."[8]

08.6 / A RAISED PLATFORM IN THE CENTER OF THE COURTYARD CAN BE USED AS A STAGE FOR CONCERTS.

08.7 / THE LABORATORIES ARE DESIGNED FOR CHANGE.

08.8

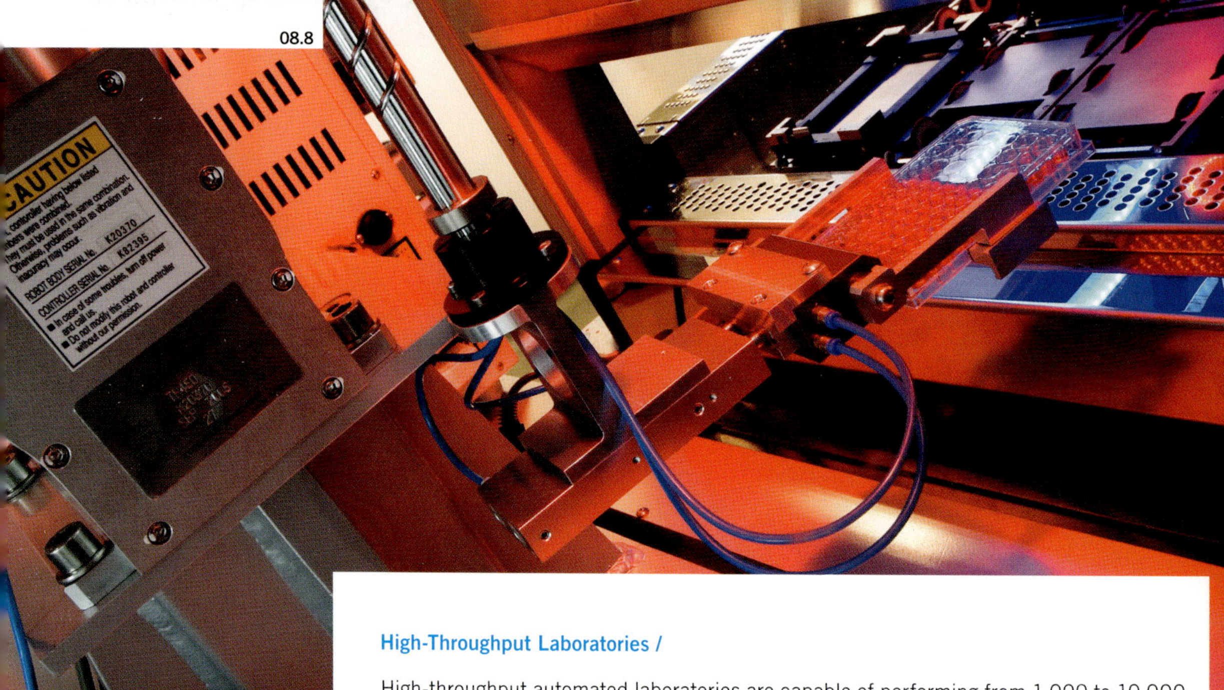

High-Throughput Laboratories /

High-throughput automated laboratories are capable of performing from 1,000 to 10,000 more tests per day than conventional labs operating just five years ago.

The operating system for the high-throughput laboratory enables secure access worldwide, allows tests to be carried out in a flexible manner, schedules and performs numerous tests on a routine basis, and deposits test results into large databases. **The operating system permits scientists to connect to the high-throughput laboratory by way of the Internet or secure intranets.** A set of process-control tools are then used to program and manage all the necessary steps for the design of tests, documentation of samples, submission of samples, and the analysis of data. **An extensive collection of samples can be achieved through a concerted worldwide effort.**

High-throughput laboratories and database networks will play a critical role in preventing, deterring, and responding to acts of bioterrorism or a viral outbreak anywhere in the world. They will facilitate efforts taking place in public health, agricultural, emergency response, law enforcement, and intelligence and national-security communities.

Case Study for High-Throughput Research /

Project: CDC National Center for Environmental Health (Building 110)

Gross square feet: 156,000

Construction cost: $60 million

Year completed: 2005

On September 11, 2001 from 9 to 10 am I was at the CDC's Chamblee Campus presenting the final schematic design for its new research building for the National Center for Environmental Health. During that same time the horrific tragedy at the World Trade Centers and the Pentagon were occurring. Messages were being brought into the conference room while I was presenting. With the faces I was looking at I thought the building design was not meeting their expectations. That was not the case—they were focused on much bigger issues. Five days later Tommy Thompson, head of Human and Health Services at the time, asked us to refocus our solution to address high throughput. The overall objective was to allow researchers to quickly change the entire building in

08.8 / HIGH-THROUGHPUT LABORATORIES ARE FULLY AUTOMATED.

08.9 / THE CDC NATIONAL CENTER FOR ENVIRONMENTAL HEALTH.

an emergency from multiple-research programs to just one project. The following pages document the solution and how we got there. This is a project I am very proud of because our design team thought outside the box to come up with new solutions to support the research programs.

Laboratory Design /

"Building 110 has allowed unprecedented focus of scientists on major public-health challenges by providing an environment that has an intelligent and flexible space design; specialized, uninterruptible, and reliable power; simplified connection of scores of advanced instruments to specialty gases and pumping systems; reliable and stable temperature and humidity control; straightforward and workable division of space by security need; unusually expansive and critical storage space outside the working laboratory; ample conference rooms positioned near lab space; daylight penetration to literally every laboratory room; and visually pleasant internal and external features," said James L Pirkle, deputy director for science at the CDC National Center for Environmental Health.

Background /

Building 110 is a multi-disciplinary, automated high-throughput laboratory facility designed to provide critical research, analytical, and related scientific services key to the efforts of public health, agricultural, emergency response, law enforcement, and intelligence and national-security communities. The building's design responds to aggressive technical, aesthetic, and user-satisfaction goals driven by heightened security concerns and emergency response, and takes the researchers' previous work environment into consideration.

Containing high-throughput automated laboratories, the facility is capable of performing from 1,000 to 10,000 tests per day. Compared to manual laboratories, these are capable of providing from 100 to 1,000-fold improvement in sample processing speeds and volumes. Such automated laboratories must be able to operate 24 hours a day, seven days a week, requiring the efficient management of a continuous flow of supplies and samples while maintaining maximum quality control efficacy on an ongoing basis.

08.9

In essence, Building 110 has been designed to fill a key role in protecting the health, safety, and welfare of national and international populations in response to any number and type of chemically related environmental situations, emergencies, and threats for decades to come. Because of the critical and highly dynamic nature of the potential challenges to be faced, the importance of successfully meeting the facility's aggressive and complex design goals would be difficult to overstate.

Flexibility and the Ability to be Quickly Reconfigured /

The design surpasses current expectations of lab flexibility, adaptability, and the ability to be quickly reconfigured—any and all spaces can be completely reconfigured within 24 hours, taking approximately 10 percent of the time needed to reconfigure traditional lab spaces.

Representative innovative solutions making this quick reconfiguration possible include:

Locating all services including wet columns, overhead service carriers, and ceiling-utility drops at the equipment-intensive building perimeter allows for the unimpeded reconfiguration of laboratory areas. A particularly innovative aspect is that perimeter-zone design allows for the addition or removal of gypsum walls without tearing down carriers or drops.

Casework is comprised of movable tables and mobile-base cabinets, providing multiple reconfiguration options appropriate to the specific needs of each research activity. Overhead shelving and cabinets can be removed or relocated to provide sufficient vertical spaces for taller equipment.

Safety zones are graphically marked on lab floors, ensuring research space reconfiguration does not defeat code requirements, thereby accelerating the reconfiguration process. Main Street lab corridors contain safety stations and showers, creating additional potential for the reconfiguration of lab space.

Interstitial floors, created above each main floor, are creatively organized to accommodate large pieces of equipment and supplies, increasing flexibility due to the elimination of equipment that would otherwise hinder the reconfiguration of the lab space.

Another extremely important aspect of the building design is the location of horizontal-distribution mechanical,

electrical, and plumbing (MEP) systems on each interstitial floor, a feature that allows MEP systems maintenance outside of the lab environment and therefore eliminates interference with research. The MEP systems are designed in a modular fashion.

The overhead service carrier is the "heart" of the lab, pumping life for the efficient and flexible functioning of these spaces. To achieve a free floor space, all utilities and services are available from the ceiling and wall.

The exterior zones of the building, which are more equipment-intensive, have overhead service carriers on a modular basis. The interior zones have vertical service drops that are able to serve a wide surface area on the floor.

Promoting Staff Communication and Interaction While Creating a Great Place to Work /

The design not only features numerous attributes capable of attracting and retaining world-class researchers and technical staff, it also actively promotes ongoing open communication through interpersonal and interdisciplinary visibility and contact between on-site researchers. Further, the design includes systems for rapid and secure communications between Building 110 and other key laboratories.

Examples of these design attributes include:

Dramatic and aesthetically intriguing entry and atrium areas with balconies, furnishings, and other accoutrements encouraging both formal and informal staff interaction.

High-tech conference rooms that allow CDC personnel to communicate remotely via audio and video in both Internet and secure intranet systems environments.

Meeting rooms imbued with interior design elements focused on user-friendliness and equipped with communications devices promoting expression of creative thought.

Numerous informal break areas.

Comfortable, emotionally uplifting, and glare-controlled daylight that penetrates 30 feet into the building.

Dramatic, 16-foot-high sloped ceilings in labs that not only allow significant daylight to penetrate into the building's core areas but also create a spectacular sense of light and openness in the labs themselves.

Seemingly omnipresent visual connections to the outdoors through multiple "view paths" that allow 90 percent of all building occupants an outside view.

Interior colors and textures that defy the sense of sterility typically expected of laboratory environments.

"The large, tasteful atrium immediately conveys a visual sense of importance and purposefulness to all employees and visitors as they enter the building.

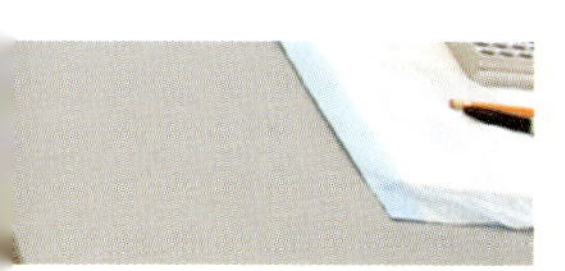

08.10 / THIS HIGH-THROUGHPUT AUTOMATED LABORATORY IS CAPABLE OF PERFORMING UP TO 10,000 TESTS EACH DAY.

08.11 / A RESEARCHER RECONFIGURES HIS LAB SPACE.

08.12

The skillful use of environmentally friendly architecture designs in all parts of the building, from energy-efficient lighting to conserving rain water for irrigation of plants around the building, reinforces CDC commitment to the importance of the environment in public health and our need to conserve precious resources for future generations," said Eric J. Sampson, director of division of laboratory sciences.

Most lab buildings have major storage issues. The interstitial floors above the office spaces, which have minimal MEP services, allow for storage of extra supplies or decontaminated instruments temporarily. Building 110 is highly equipment-intensive, leaving the integrity of the space plan vulnerable when new or additional equipment is acquired. This interstitial storage space allows true flexibility due to the absence of reconfiguration-hindering equipment. The space gives the ability to purchase and store single-lot project supplies, making it more cost effective and analytically stable. Easy access to storage supplies, rather than storing in ancillary buildings or off-site, is a great convenience for the researchers.

Translational Research /

Translational research is the direct link between the laboratory and the patient's bedside. Often called "bench-to-bedside," it is the transfer of knowledge into clinical practice, which travels from bench to bedside and back. It can also refer broadly to the development of new technologies whereby early patient testing is emphasized.

Today's medicine is increasingly a patient-driven research process.

Translational research takes place between a fundamental discovery and the application of that discovery to medicine. It provides a much-needed bridge of communication between highly specialized research scientists and physicians. Researchers, who work in labs with microscopes and tissue samples, and physicians, who directly interact with patients, speak very different languages. Cutting-edge research must be delivered to physicians in a useable format. It is also essential that physicians pass on information to researchers.

08.13

In addition to being the application of basic scientific discoveries to clinical research, translational research is also the generation of scientific questions based on observations made in humans. This dual meaning is significant because if research is to be effective, it must avoid routes that are too linear. On the one hand, animal models used at the preclinical stage offer a mere flawed reproduction of the complexities involved in human diseases; on the other hand, many basic questions stem from epidemiological and clinical observations.

We must increasingly streamline the process of making basic biomedical discoveries, whether in the form of new diagnostics, therapeutics, or even healthcare guidelines. But a gap exists between the point at which research universities and institutes typically terminate the discovery process and the point at which pharmaceutical companies typically initiate drug development. Pharmaceutical companies tend to be interested in drug lead compounds rather than in early-discovery research.

The challenge is to make these gems "presentable" to industrial partners prepared to take risks. This "proof of concept," an interim phase of 18 to 24 months when the process of presenting projects can be accelerated and take on a more tangible form, is a critical weakness in the relationship between the public and private sectors.

Loads of public financing and private investment has been poured into US biomedical research. In 2003 the various National Institutes of Health put forward an idea called the NIH Roadmap. By facilitating information exchange between basic and clinical research, the roadmap has established vertical links enabling researchers to test hypotheses on potential preventive measures and therapies.

08.12 / AN IMPORTANT ASPECT OF THE BUILDING DESIGN IS THE LOCATION OF THE MECHANICAL, ELECTRICAL, AND PLUMBING (MEP) SYSTEMS.

08.13 / THE DESIGN PROMOTES ONGOING OPEN COMMUNICATION.

Opportunities /

Scientists are increasingly aware that this bench-to-bedside approach to research is really a two-way street. Scientists provide new tools to clinicians so that they can be tested on patients, and clinical researchers make novel observations about the nature and progression of disease that often stimulate basic investigations. Recognizing the gap between researcher and physician, the National Institute of Environmental Health Sciences (NIEHS) developed a series of translational research programs that share the following three objectives:

1. Improve the understanding of how physical and social environmental factors affect human health.
2. Develop better means of preventing environmentally related health problems.
3. Promote partnerships among scientists, health care providers, and community members.[9]

Examples of translational research programs include stem cell research, cancer immunotherapy, clinical pharmacology, immunology, gene therapy, infectious diseases, neurobehavioral development, and molecular medicine.

"The NIEHS defines translational research as the conversion of findings from basic, clinical or epidemiological environmental health-science research into information, resources, or tools that can be applied by health care providers and community residents to improve public health outcomes in at-risk neighborhoods. In addition, the NIEHS gives special attention to insure that the information is culturally relevant and understandable."[10]

"Gene therapy is the focus of research for a large group devising ways to reawaken the expression of silenced genes that can inhibit cancer growth. Testing is occurring for several different drugs to protect people against prostate, breast, and colon cancer. There are studies of retinoids to prevent tobacco-initiated cancers. A way to identify people in large populations at risk for certain cancers who would benefit from cancer screening and chemoprevention ... It is now possible to imagine designing a specific prescription for each patient wherein you would treat just exactly those abnormalities that occurred in their cancer ... There is translational research at all different sites—breast cancer, gastrointestinal cancer, lung cancer, and prostate cancer. There is also translational research in early detection, diagnosis, prevention, and treatment."[11]

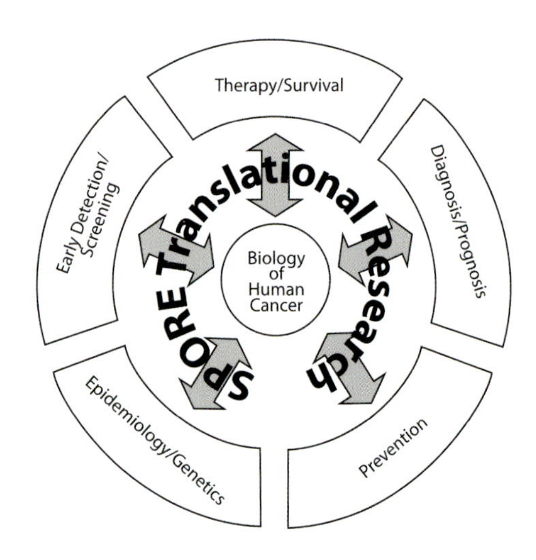

08.14 / THE TRANSLATIONAL-RESEARCH MODEL

A sharp drop in heart attack deaths in more than a dozen countries coincides with global efforts to make sure patients receive proven treatments. A study of 44,372 patients in North America, Europe, and South America from 1999 to 2000 found that deaths, heart failure, and cardiogenic shock (when the heart goes into shock and loses pumping power) all declined in patients hospitalized for heart attacks or for life-threatening chest pain.[12]

In 2001, the American Heart Association launched an effort to encourage doctors to follow guidelines for heart care based on the latest scientific evidence. Two years ago, the agency that pays for Medicare began docking hospitals part of their repayment, now up to 2 percent, if they didn't report what percentage of their patients were getting certain guideline-related heart therapies. The European Society of Cardiology has taken more informal steps to achieve the same goals. During the study, hospital deaths fell 18 percent in patients with the most severe form of heart attack, the study shows. There also was a significant decline in rates of stroke and heart attack six months after the initial hospitalization, researchers say.[13]

As the study unfolded, more doctors worldwide were adopting state-of-the-art approaches to therapy. For instance, in 1999 only one-third of heart attack patients were treated with angioplasty, though angioplasty cuts the risk of death by 40 percent more than the clot-busting drugs. By 2005, 64 percent of all heart attack patients received angioplasty, the researchers reported in *The Journal of the American Medical Association*. The study reflects the relative success of programs meant to bridge gaps between research and patient care.[14]

I had the opportunity to meet with Dr. David S. Stephens, vice president for research at Emory University School of Medicine, to discuss his thoughts on where translational research is today and where it is headed over the next 10 years. He said that translational research is often seen to be a process that goes from "bench to patient." Dr. Stephens referred to this as T1 research, which is comparable to phase 1 clinical trials. But this is only part of the story. "Bedside to clinical practice" is the next area (T2 research), which looks at safety and efficacy at the level of phases 2 and 3 clinical trials. The information is taken to the FDA to be developed together with the research team.

The third area is "clinical practice to policy" at the T3 and T4 research levels. This involves community participation with feedback that might help to drive the research. **With clinical practice to policy there seems to be more flexibility and an opportunity to fast track some research with the FDA.** However, once the compound is established, it must not change and if it does then a new study is required.

There is now a broader definition of research. Healthcare research is now more focused on **"clinical effectiveness research," which tries to be more time and cost effective.** The new research focus is to try to influence the clinical trials. In the USA much of the money that goes into healthcare is wasted, even though more is spent on healthcare than that of any other country by more than 50 percent. Dr. Stephens explained that administration alone accounts for one fourth of the costs attributed to healthcare.

We also had some time to discuss Dr. Stephens' thoughts on personalized medicine. I asked him how he saw personalized medicine impacting translational research over the next 10 years. He said that the first concern is to be able to provide affordable personalized medicine for as large a population as possible. He explained that genetics will determine the drugs we require for specific ailments and their quantities. There will need to be an emphasis on personalized care and that we will need to look at public health and at the overall population to make improvements in our healthcare. Public-health-related solutions may conflict with personalized medicine, but ideally there will be improvements in both to help improve the overall quality of healthcare.

Virtual-Reality Applications /

Virtual-reality (VR) applications are poised to transform both research and treatment in drug addiction. Much of the following information on VR was developed with Patrick Bordnick, a researcher at the University of Houston, Graduate College of Social Work.

New VR applications have been developed and tested in people addicted to nicotine, alcohol, marijuana, and cocaine. The goal of VR in addictions is to create immersive experiences involving complex drug-related stimuli and to provide standardized cue-reactivity and

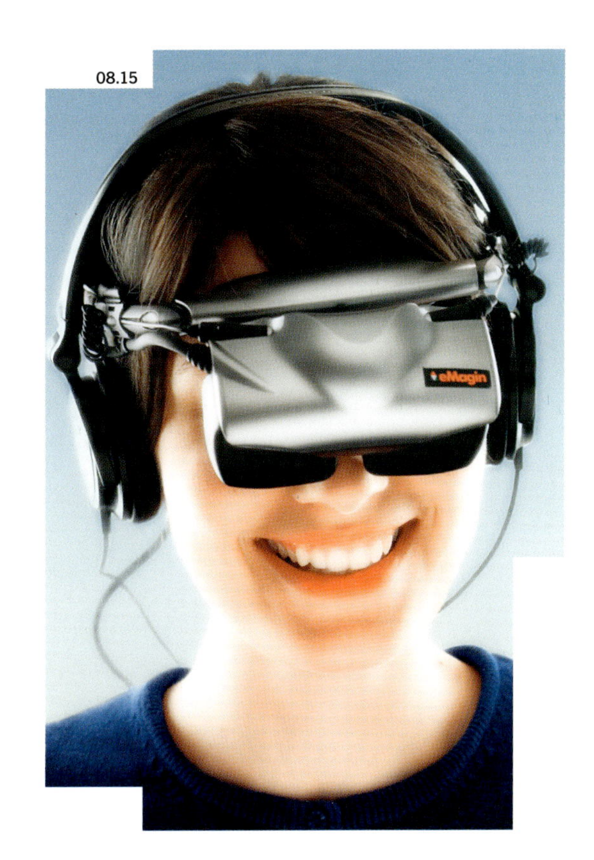

08.15 / VIRTUAL-REALITY TOOLS INCLUDE THE USE OF MULTIPLE-SENSORY SYSTEMS.

08.16

craving-assessment systems for researchers and clinicians. VR can provide valid settings in which to teach relapse-prevention strategies in real-time. VR tools are expanding to include multiple-sensory systems in controlled clinical and research settings. The ability to bring complex stimuli into imaging allows the design of new experiments and the testing of novel agents purported to change addictive behavior.

The US Substance Abuse and Mental Health Services Administration 2003 survey found that nearly 22 million people over the age of 12 met the criteria for substance abuse or dependence. The number of users of alcohol, tobacco, and illegal drugs, respectively, were estimated at 119 million, 71.5 million, and 19.5 million. In the 12 months prior to the survey, 22.8 million people in the USA had sought specialized treatment for drug or alcohol problems. Given the high number of individuals seeking treatment, often with multiple admissions and relapses, new assessment and treatment approaches are warranted.

A diagnosis of substance abuse or dependence requires assessment of behavioral, biological, and psychological symptoms by a trained clinician. Substance abuse is classified as a dysfunctional use of substance that leads to significant problems in social, medical, and legal areas related to use. Examples might include the continued use of marijuana after repeated drug arrests, or repeated alcohol consumption after repeated drunk-driving arrests. Substance dependence builds upon the abuse criteria, and adds the repeated use of a substance that leads to the development of tolerance and withdrawal symptoms. Diagnostic criteria are listed for the following specific substances: cocaine, alcohol, nicotine, marijuana, amphetamines, hallucinogens, opioids, caffeine, phencyclidine, sedatives, inhalants, and hypnotics.[15]

Cue-Reactivity/Exposure /

Cue-reactivity assessment involves exposing substance users to cues/triggers in a controlled setting and measuring subjective craving and physiological reactivity. In traditional laboratory studies, participants are exposed to both drug cues and neutral cues (non-drug related stimuli), and assessment of craving intensity for the drug and autonomic physiological reactivity are recorded and compared across cues. Historically, cue-reactivity studies have relied upon imaginal exposure (imagery scripts), role-playing (e.g. social situations), still photographs, video/audio, and drug paraphernalia (tactile stimuli).

Imaginal exposure involves having the participant listen to a script describing a drug-use scenario—all other cues are presented in a controlled fashion allowing inference

08.17

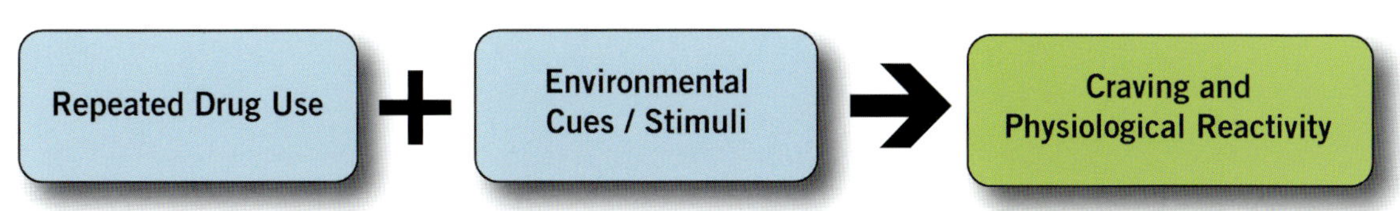

about the relative strength of the cue within an individual participant. In order to simulate drug-using situations or drug-sales offers, role playing is used to create these interactions. During such role plays, the researcher or clinician plays the role of a drug-using friend or dealer and makes an offer to use a drug with or sell a drug to the participant.

Typical visual drug and neutral cues are still photographs of either the drugs of abuse or neutral scenes (e.g. nature or other non-drug related items). Video cues often depict persons using abused substances or containing scenes of drug-using environments such as a crack house or a bar. Lastly, drug paraphernalia (tactile cues) are handled by participants to simulate drug use. Common tactile-cues items include beer bottles, crack pipes, bongs, cigarettes, and lighters. It is worth noting that cues/stimuli used in cue-reactivity studies vary across study sites and experimental labs. Broadly, traditional laboratory clinical research has produced results indicating that exposure to drug cues elicits both increased physiological arousal and subjective craving when compared to neutral cues.

While the results from these studies suggested a role for cue-reactivity/exposure in substance-abuse treatment, there have been difficulties translating findings from traditional studies into effective treatment programs. Several methodological problems may be the root of these difficulties. Commonly, cues utilized in research settings are specific, austere, and may lack complexity. These attributes allow researchers to thoroughly study specific questions and objects, though cues rarely occur exactly these ways in the real world. To aid in the application of effective cue-exposure techniques for drug treatment, there is a need to develop and test more complex cues that include visual, audio, tactile, and olfactory senses. Developing a VR cue-reactivity system can provide exposure to complex cues (such as combinations of social situations and interactions, affective experiences, olfactory, and physical stimuli) and may well improve generalization of skills and exposure back to real-world settings. Given current remission rates after addictions treatment, the development of new approaches that utilize ecologically valid cue exposure systems are clearly warranted. In the next sections, the development process, implementation, and clinical testing of VR-based cue-reactivity systems for nicotine, alcohol, marijuana, and crack cocaine are discussed.

08.16 / GRADUATE STUDENTS CREATE SIMULATION MODELS TO TEST SPECIFIC ENVIRONMENTS.

08.17 / CUE ELICITED REACTIVITY BASED ON RESPONDENT CONDITIONING THEORY. OVER TIME, DRUG USE IS PAIRED WITH STIMULI IN THE USERS' ENVIRONMENT AND THESE STIMULI AFTER REPEATED PAIRING CAN ELICIT REACTIVITY WITHOUT THE DRUG BEING PRESENT.

08.18 / SAMPLE SIMULATION ENVIRONMENT WITH RESEARCHER USING HEADSETS.

08.18

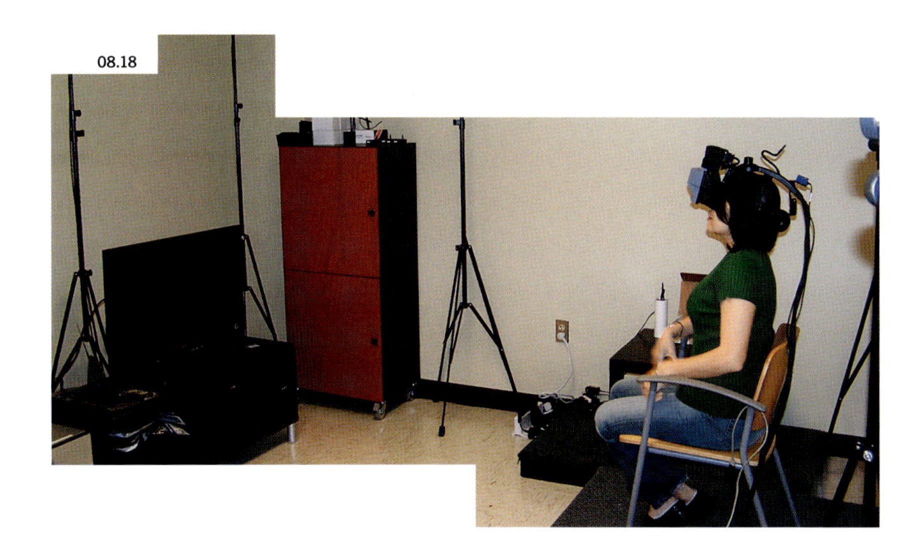

08.19

Aims of VR Addiction Research /

Initially the development of VR cue-reactivity systems sought to design environments that simulated both inanimate—burning cigarettes, coffee pots, beer bottles—and animate cues— engaging social interactions. The initial environments also contained neutral cues/ stimuli allowing comparison between the cue-rich experience and neutral spaces. VR cue systems provide immersive experiences using visual, auditory, olfactory, and tactile cues.

Participants are isolated in the VR environment with a head-mounted display and magnetic tracking system to minimize outside distractions. Social interactions are simulated using video that creates realistic settings while at the same time allowing for a safe and controlled environment, and enables professionals to communicate with participants in real-time. Lastly, VR offers the ability to duplicate and repeat infinitely complex cue situations controlled by the researcher or clinician.

In order to go beyond traditional cue-reactivity methods and treatment programs, a VR solution must provide acceptable cues with usable social interactions, freeing the professionals from role-playing and other ineffective duties. Participants also must become immersed in VR environments; they must act, react, and interact as if they are actually present in the scenario being produced. This level of immersion will aid in providing valid and reliable VR environments for use in the assessment and treatment of drug and alcohol dependency.

VR Nicotine Cue-Reactivity Assessment System /

The first such VR nicotine cue-reactivity assessment system (VR-NCRAS) was developed using an STTR collaboration grant funded by the National Institute on Drug Abuse (NIDA). In the development phase, two virtual-smoking cue environments and one neutral environment were created. The first virtual-smoking environment consisted of a room with various smoking paraphernalia (cigarette packages, ashtrays, burning cigarettes, and alcoholic beverages) on tables.

The second smoking virtual environment consists of a party setting. In this module, people are seen smoking, drinking, and interacting with the participant. After engaging in limited

08.20

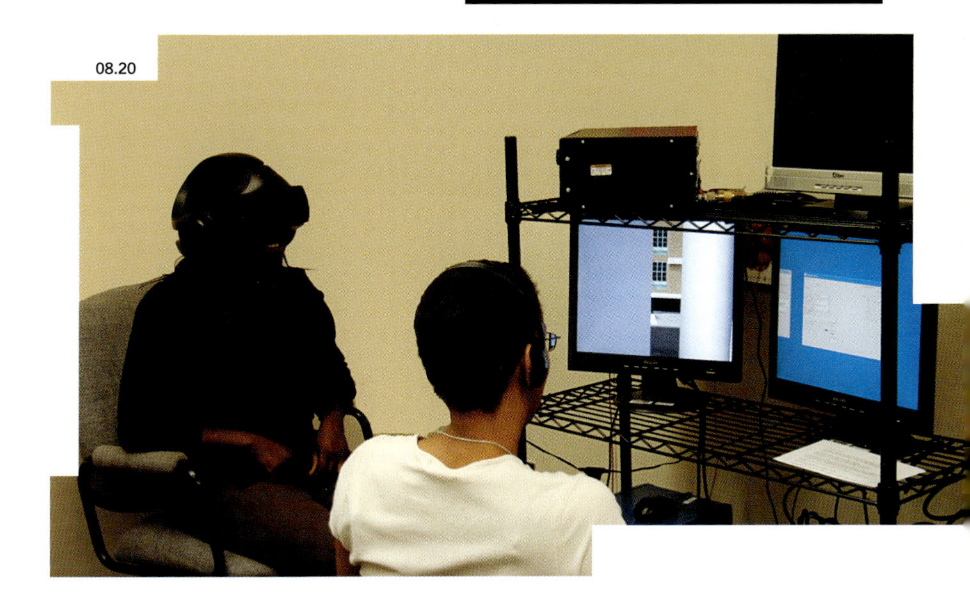

conversations, the participant is offered a variety of different cigarettes by VR characters. The smoking party and social interactions represent an advance from traditional role-playing in cue-reactivity research, by allowing participants to interact with video images of people who are smoking and drinking in the virtual world and freeing the professional from unrealistic role playing duties. Finding a neutral environment to use with nicotine addicts proved more difficult than originally anticipated. This may have been related to the association between smoking and relaxation.

Ultimately, the neutral environment that demonstrated the no-craving response in trials consisted of an art gallery containing aquarium scenes that play when subjects approach along with accompanying audio (bubble sounds). Such scenes immerse participants in an environment that simulates being underwater. To complete the design, representations of all the rooms in the house, including the room containing the party, are programmed to allow control by researchers. In addition, timed audio clips provide sound effects for various objects (percolating coffee pot, ice clinking in glasses, doors opening) and smoking for burning cigarettes were created by using available 3D tools. Smoke was added to the burning cigarettes using a particle generation system. Interactions with the smoking paraphernalia and the smoking social interactions are also controlled by the researcher.

The VR-NCRAS was tested in two controlled pilot trials with nicotine-dependent smokers. Smokers in both trials experienced increases in craving in smoking rooms compared to neutral rooms. On average craving increased over 110 percent between neutral and smoking rooms. Participants rated the VR environments post session as creating a feeling of presence and immersion during the experience. Interestingly, during the trial many of the participants actually reached for cigarettes in the virtual ashtray and packs that were on a virtual table.

This provides further support for the ability to be able to generalize into real-world settings. After this successful development and testing phase, the VR-NCRAS was expanded to six additional VR environments into a treatment study for smoking cessation. In this study, VR will be integrated into a treatment program to allow therapists to expose nicotine-dependent smokers to high-risk situations, while teaching relapse-prevention and coping skills. This environment was the first controlled demonstration of craving in VR.

08.19 / ACTUAL COMPUTER-SIMULATION MODEL CONTAINING INANIMATE AND ANIMATE CUES.

08.20 / VIRTUAL-SMOKING ENVIRONMENT.

VR Crack House /

The VR crack house was developed by Dr. Mike Saladin, Medical University of South Carolina, and Virtually Better, Inc., with funding from NIDA. The goal of this project was to construct a demonstration unit to evaluate craving in non-treatment-seeking crack-cocaine addicts. Unique features of this research included the development of cue content that specifically addressed crack-cocaine addiction.

Significant research was needed to understand the stimuli associated with this addiction.

Georgia's Dekalb County Police Department provided access to areas that had been used as crack houses and offered an overview of what life is like inside them. Dr. Saladin interviewed people in treatment, who provided vivid descriptions of their lives as crack-cocaine addicts, and translated these personal accounts into specifications that helped direct the development of the VR environment.

Ultimately, the virtual crack house included areas where sex is traded for drugs (audio only), someone acts as a lookout, people are "sleeping off" a high, a drug deal takes place, people are using the drug, and offers are made to sell crack cocaine. The final scenario occurs in a darkened room while the house is being raided by police. The control condition was essentially the same as that found in the nicotine environment (underwater scenes). Because nicotine addiction is also common among crack-cocaine addicts it was necessary to avoid the initiation of nicotine craving.

Results from the first trial indicated approximately a 400 percent increase in craving among the subjects who entered the experiment. The participants were interviewed about their experiences and based upon that information a phase-2 application has been submitted to develop additional environments in a treatment setting. Additionally, many mental-health providers have been able to experience a crack house in VR without risking their personal safety.

VR Alcohol Cue-Reactivity Assessment System /

Traditional alcohol cue-reactivity research, similar to nicotine research, had been limited by austere cues and a lack of complex stimuli. However, the VR alcohol cue-reactivity assessment system (VR-ACRAS) has been developed by Virtually Better, Inc. in collaboration with Dr. Bordnick at the University of Georgia, with funding from an STTR grant provided by the National Institute on Alcohol Abuse and Alcoholism (NIAAA).

08.21 / VIRTUAL-SMOKING ENVIRONMENT IN A PARTY SETTING.

Building upon the initial success of the VR nicotine cue-reactivity, the VR-ACRAS includes the following:

- The addition of an affect-based social interaction cue, designed to create a negative-affect state (negative affect is often associated with alcohol relapse and use)
- Various social interactions, which provide opportunities to test subtle verbal drinking cues
- Integrated pop-up rating-scales assessment, and
- The incorporation of a novel scent-cue system for specific and ambient olfactory cues.

The VR-ACRAS consists of four alcohol environments and two neutral environments. The VR alcohol environments are:

1. A bar setting in which the participant watches a virtual bartender serve a drink to a customer and then is asked to order their preferred drink
2. A kitchen area where alcoholic drinks, bottles, cases of alcohol, blenders, drink mixers, and party supplies are on display
3. A home office setting in which two people drinking alcohol become engaged in a heated argument (negative-affect scenario) while the participant watches and eventually is drawn into the dialogue
4. A party setting consisting of a living room, where people are drinking and offer the participant a drink of their choice, and an outside patio deck where people are smoking cigarettes and drinking.

Olfactory Cues /

A novel method for delivering olfactory cues into VR has been developed to further the sense of immersion and realism. Olfactory scent cues are delivered into the VR setting using the Scent Palette™ (Envirodine Studios/ Virtually Better, 2004). The Scent Palette™ is a USB device that provides both ambient and specific scent cues triggered by the VR software. Ambient scent cues may include cologne, cigarette smoke, food, and mold/mildew. Scents can provide general cues linked to places such as a bar or crack house, or specific scents can be linked to individual objects such as a bottle of beer, or a lit cigarette. Currently, specific scents being tested in the VR-ACRA include lime, pizza, cologne, and cigarette smoke. Pine is being tested to simulate the scent of trees surrounding an outside deck area. We envision inclusion of scent cues in all future substance-abuse VR applications to enhance immersion in cue-reactivity environments.

Future directions /

VR is beginning to emerge as a potential substance-abuse treatment and assessment tool for researchers and clinicians. VR cue-reactivity has been shown to be a viable method to assess craving and physiological reactivity in nicotine-dependent smokers. Data from clinical tests of VR cue-reactivity with crack-cocaine, cannabis, and alcohol will also be available soon.

This technology has the ability to offer several advantages over traditional assessment and treatment methods. First, VR participants are immersed in visual, auditory, olfactory, and tactile environments without outside distractions. Second, video images allow the participants to interact with people instead of role-playing or using computer-animated avatars. Third, researchers gain a level of safety, control, predictability, and repeatability in VR-exposure situations that is not available in the real world. Fourth, researchers maintain constant communication with participants through audio links that offer opportunities to teach relapse prevention, drug refusal, and coping skills in real time. Fifth, access to a set of repeatable, complex cue situations (e.g., social interactions) controlled by the researcher, used for exposure and skills acquisition will allow advances in addictions research. Sixth, the recent addition of a novel computer-controlled USB device to simulate various scents (olfactory cues) during VR provides additional immersion and realism.

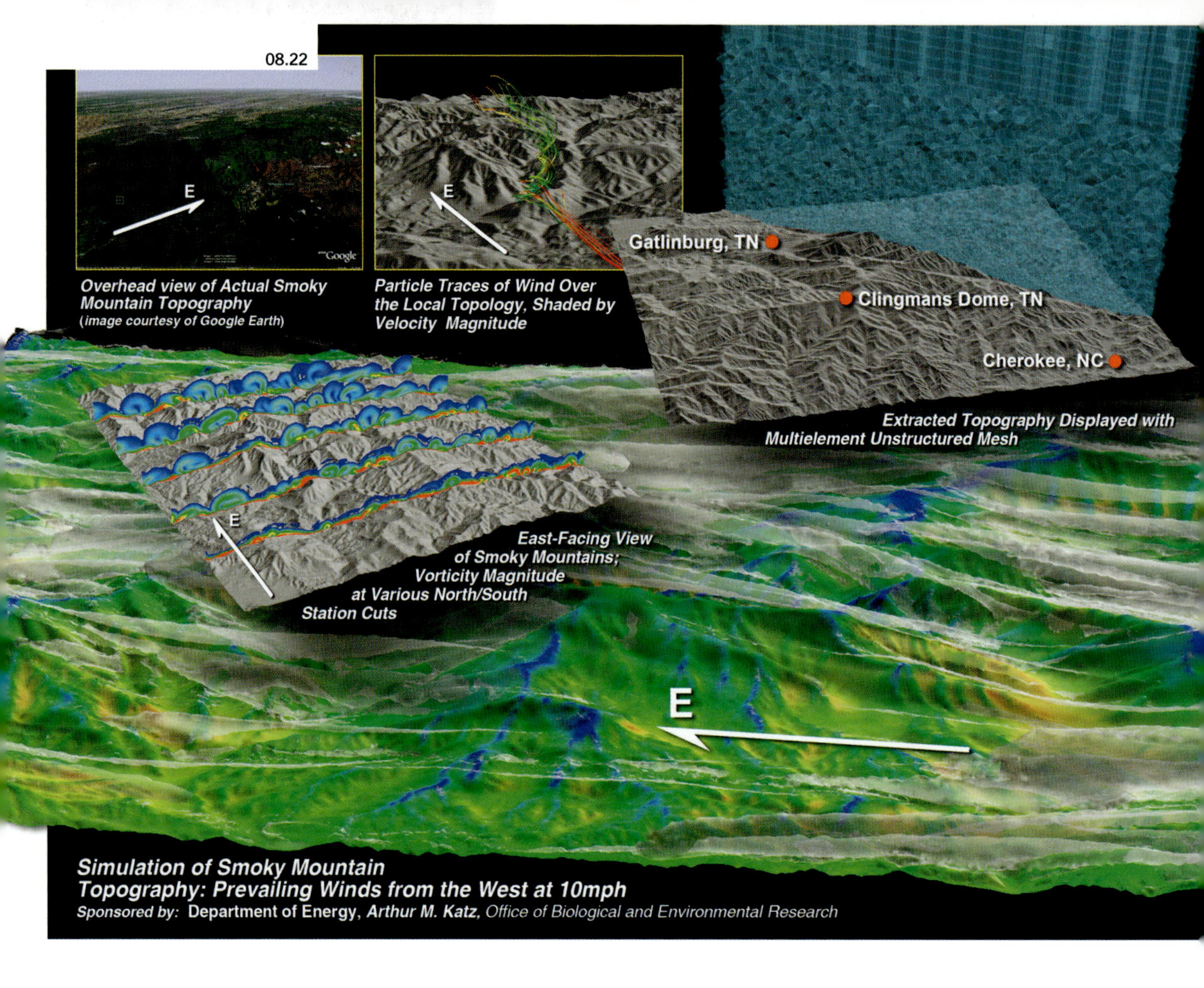

Simulation of Smoky Mountain Topography: Prevailing Winds from the West at 10mph
Sponsored by: **Department of Energy, Arthur M. Katz,** *Office of Biological and Environmental Research*

Computational Engineering /

Computational engineering is an emerging multidiscipline that solves complex, practical, real-world engineering analysis and design problems using advanced computer simulations based on physical and mathematical models. The University of Tennessee at Chattanooga is developing one of the leading programs in the USA. Some research studies include:

- Simulation of wind patterns in the Smoky Mountains
- Release of containment in an urban environment
- Air-pollution analysis in an urban environment
- Ship motions in incident waves
- Aircraft maneuvering
- Simulation of drag reduction for large truck to improve gas efficiency.

The new Center for Nanoscale Materials Research at Argonne National Laboratory outside Chicago allows clean-room laboratories to study various nanoscale materials, and is an addition to the linear accelerator. Since the linear accelerator is larger than a baseball field it can be difficult for people to communicate and work with each other. The world-class facility supports visiting teams from around the USA and the world and the virtual technology allows **team members located in other parts of the world to communicate in real time, both visually and verbally**, with researchers at the facility. Some call this type of research model "**telecollaboratory**."

Advances in communications, collaboration, and visualization technologies provide new ways for scientists and engineers to share and analyze complex data. The multimedia display, presentation, and interactive environment, called the Access Grid, supports large scale distributed meetings, collaborative work sessions, seminars, tutorials, and training. Wall-sized display technology, or hyper walls, consist of 10 million pixels that allow for the visualization of simulations and other digital information in great detail.

There is a main shared common space outside the clean-room area with a hyper wall that has 12 high-resolution

screens. Each screen is a 46-inch-wide high-definition television flat screen. Each screen is hooked up to a camera in a laboratory. The camera is located in a strategic position usually looking across the widest part of the laboratory to provide the most information. The camera is on 24 hours a day, seven days a week.

Viewing the various labs in this shared public space serves many benefits. First, guests are provided a virtual tour of the facility without going into any of the laboratories. This does away with the need for glass walls along the lab corridors, which in turn provides more wall space for furniture and equipment. The images can also be viewed by members of the research team who can communicate with researchers in the laboratory. The cameras also help to improve security. Another advantage of virtual tours in the public area is that the research team is not interrupted allowing them to work more efficiently.

Hyper walls are also installed in certain labs where research teams are required to share information with people outside their laboratory. The screen may show research occurring at a nanoscale with a nanodevice.

Computers are connected to the equipment to acquire data as the research is being conducted, and the data is then distributed to whoever requires it. Laptops and flat screens are located immediately adjacent to most sophisticated research equipment.

White boards are used to write ideas down and sketch up thoughts and then the information can be transferred back to the laboratory. These boards are often used as a part of teleconferencing. Kiosks can be installed to create posters generated from computer data, equipment, or white boards. A convenient area where the posters can be displayed should be considered so that the ideas and information can be shared.

A new model of research is the use of robotics from a distance to support research. When it's perfected, telesurgery could quickly become the medical norm for remote locations such as battlefields, isolated communities, and even space.

Surgical training and evaluation has traditionally been an interactive and slow process in which interns and junior residents perform operations under the supervision of a faculty surgeon. This method of training lacks any objective means of quantifying and assessing surgical skills. Economic pressures to reduce the cost of training surgeons and national limitations on resident work hours

08.22 / SIMULATION OF SMOKY MOUNTAIN TOPOGRAPHY: VIRTUAL RESEARCH ACCELERATES SCIENTIFIC DISCOVERIES AND HELPS TO SIMPLIFY BUSY AND CHALLENGING WORK SCHEDULES.

08.23 / HYPER WALLS ALLOW DIGITAL INFORMATION TO BE PRESENTED IN GREAT DETAIL.

08.24 / THE HYPER WALL ALLOWS GUESTS TO TAKE A VIRTUAL TOUR OF THE FACILITY.

have created a need for efficient methods to supplement traditional training paradigms. While surgical simulators aim to provide such training, they have limited impact as a training tool since they are generally operation-specific and cannot be broadly applied.

"Robot-assisted minimally invasive surgical systems, such as Intuitive Surgical's da Vinci, introduce new challenges to this paradigm due to its steep learning curve. However, their ability to record quantitative motion and video data opens up the possibility of creating descriptive, mathematical models to recognize and analyze surgical training and performance. These models can then be used to help evaluate and train surgeons, produce quantitative measures of surgical proficiency, automatically annotate surgical recordings, and provide data for a variety of other applications in medical informatics."[16]

Lean Management /

Private industry is beginning to re-evaluate the way it works and is attempting to be more scientific in its research process.

Some pharmaceutical companies, for example, are beginning to manage their research processes in the same way that they manage their manufacturing processes—the daily work flows are being studied, challenged, and simplified.

08.25

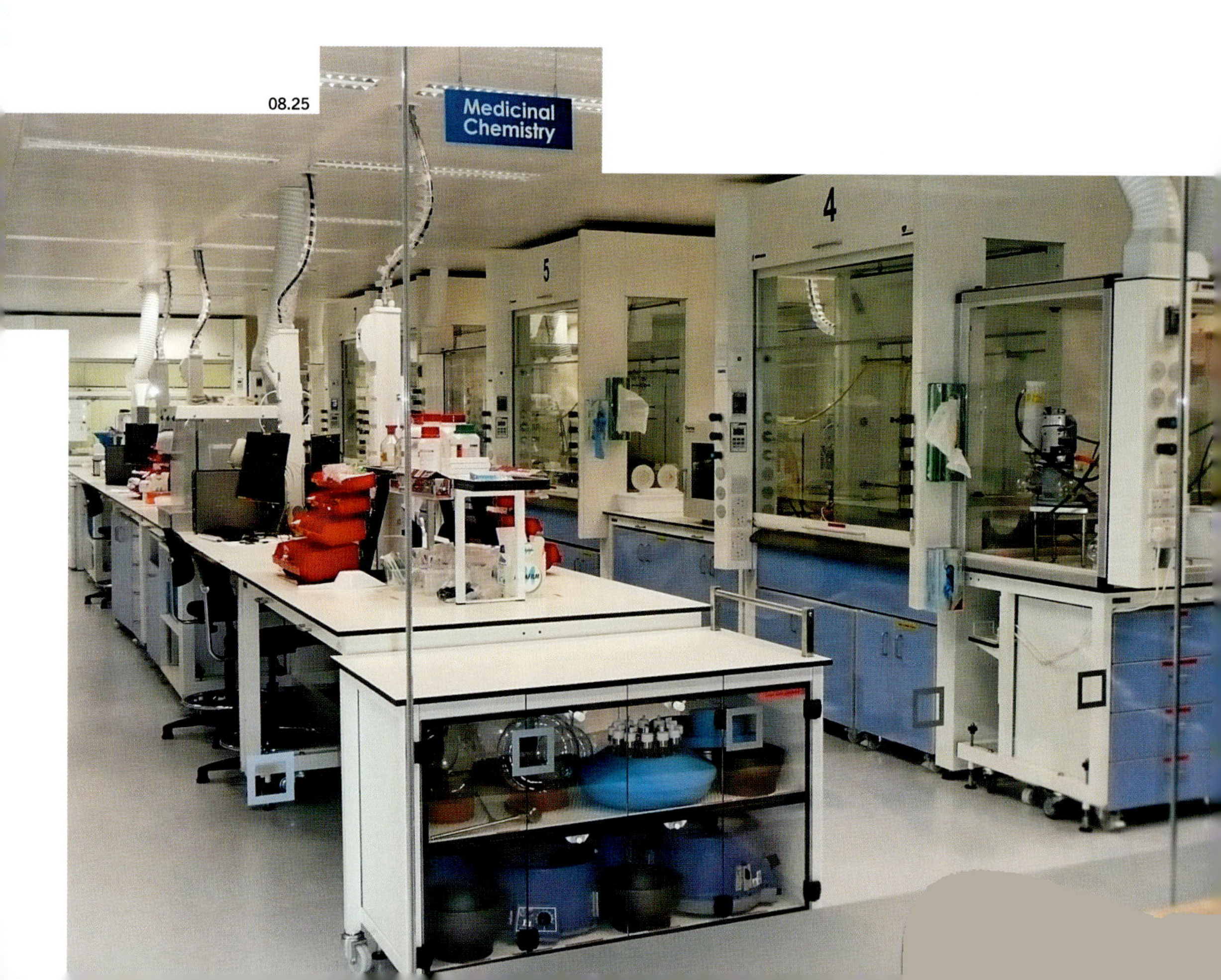

08.26

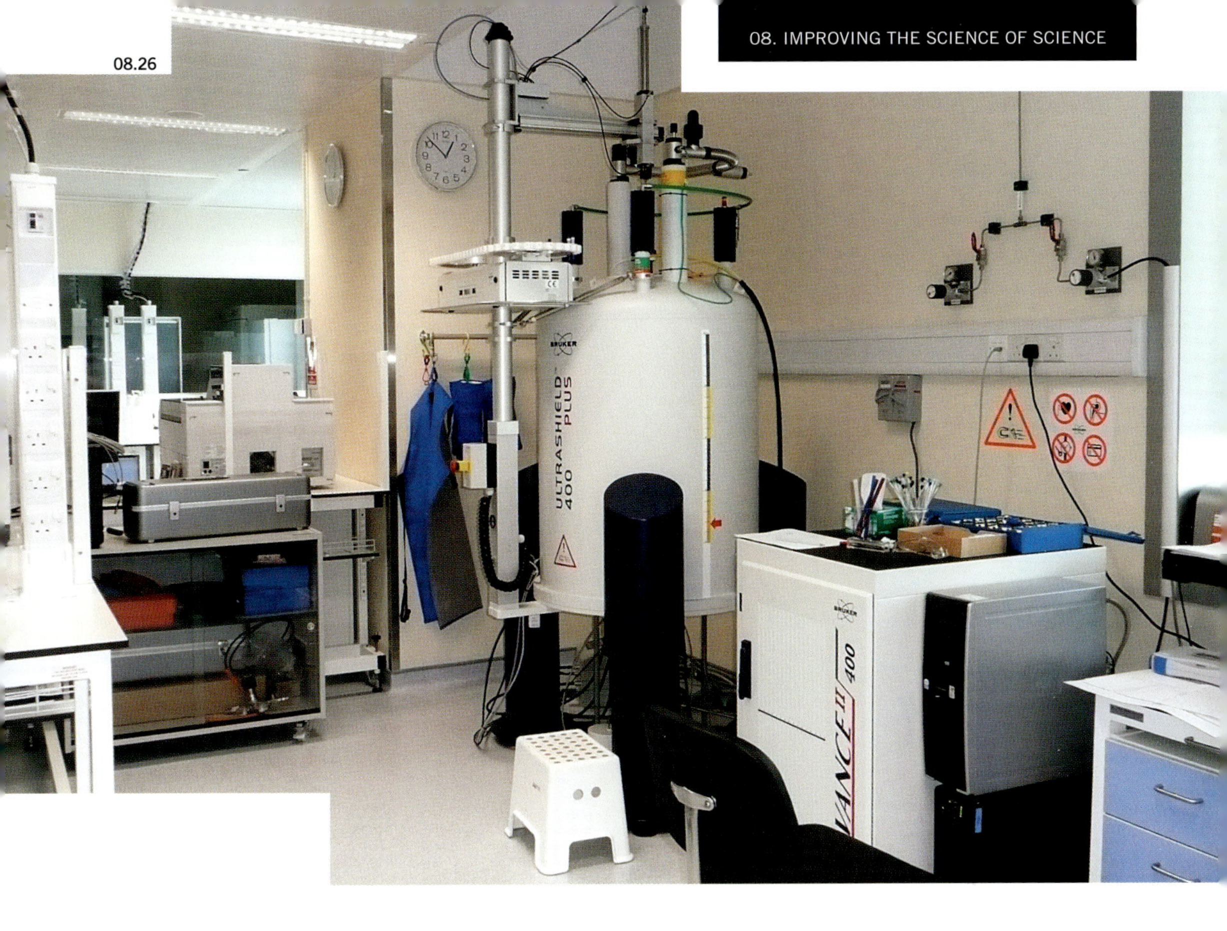

There are several variations of the lean management process that focus on improving the "science of science." The following are some key points.

Sort and throw away: The research team is asked to look over what is in their lab and determine if it is still necessary. If the supply is not necessary then it is thrown out with the intent of keeping the laboratory clean.

Store, has a home, and is labeled: What needs to be stored is determined and a place is found then labeled to help organize the lab. KanBan is a Japanese approach to managing supplies. A signboard lists all the necessary lab supplies along with the lowest acceptable amount before more supplies need to be ordered. This minimizes storage costs, wasted space, and clutter in the lab.

Shine and clear the space: The intent here is to only use the bench areas as work spaces and not for storage. Overhead shelving is often reduced or eliminated because there is little need for the storage space.

Standardize: The bench size with mobile casework is standardized to be as effective as possible. Typically 3-foot modules are used because that is the typical area required for knee space. Tables are typically 6-feet long accommodating one 3-foot-wide mobile base cabinet and knee space, or two mobile base cabinets.

Sustain: Manage and set up routines during the day and over the week to make effective use of people's time. The research team evaluates workflow, then modifies the lab to support a more effective organization. If this process is done correctly then the research team should be able to work efficiently in the same area.

08.25 / A LEAN LAB.

08.26 / NMR IN AN OPEN LEAN LAB, WHICH WAS USUALLY FOUND IN A SEPARATE ROOM IN THE OLD RESEARCH MODEL.

08.27

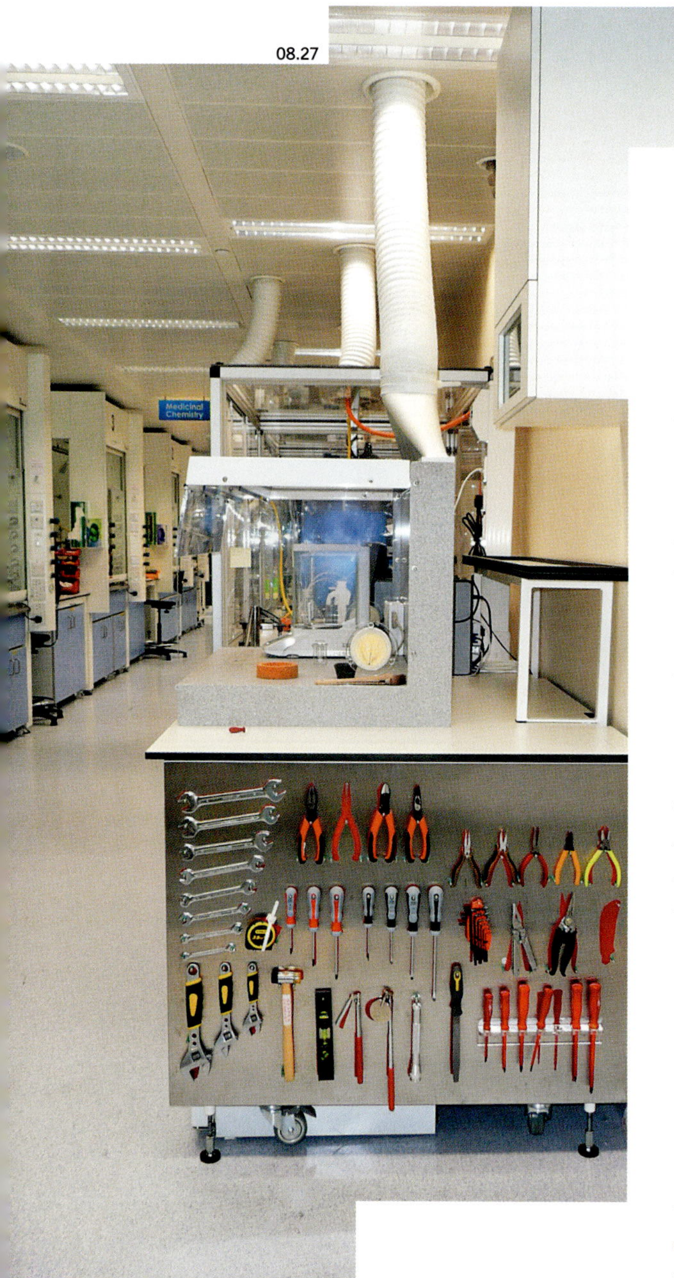

While visiting laboratories in Singapore, representatives of a multinational company shared with me how various research teams used the lean management process to develop more efficient workflow. The laboratory manager asked the team how many mobile base cabinets were required, and the initial response was 220. The laboratory manager then asked each person to justify what he or she was going to put in each cabinet. After the process was completed the team only ended up requiring 63 cabinets. After the research team had worked in the space a few weeks they commented on how easy it was to find supplies, that they had more room to work in, and that overall the space was much more effective. The team also appreciated that it was easy to modify their labs to accommodate changing research requirements.

The lean management philosophy is catching on with private companies motivated to improve research costs. Publicly funded laboratories will take longer to become more efficient. Ideally, more efficient labs should lead to smarter investments in research.

The process of research has changed significantly over the past 10 years, and will continue to evolve rapidly over the coming decades.

> As research moves forward, it is important to improve the "science of science" to reduce costs and time. Competition will become fierce.

The top research companies are redesigning their supply chain and working with vendors in very new ways. The goal is **"just-in-time supplies" to minimize storage in the building and lab. Less need for storage frees up space for the research teams to work more effectively.**

As the laboratory is being re-evaluated the focus is now on the project team and the process used to conduct research, which can change almost daily. The focus is no longer on the amount of net square feet or linear feet of bench required. The laboratory space will be flexible and adaptable enough to support the research teams.

Three types of spaces are evolving. The first is the lab zone where the research activities and scope are conducted in a safe environment. The second is the collaboration space that supports simultaneous problem solving in a team environment. The third is the private space, such as offices, that is designed to shut off communication devices so individuals can remove themselves from the team environment to quietly focus on their work for a period of time.

As lean management becomes an accepted part of the research process it will have a tremendous impact on the research, space, and the human-resource hours necessary to complete projects. Economic pressures will improve the science of science significantly over the next 10 years.

While visiting Saudi Arabia, China, and Singapore recently, I discovered that, because of limited budgets, **developing countries are already focusing on the principles of lean management**. In Saudi Arabia, I have been a part of a team responsible for designing the research centers for three new medical schools. Each campus will have warehouse storage facilities near the campus to bring supplies in each day. Supplies in the labs will be stored for up to seven days only. This means research buildings will be used more for research and less for storage. This also means less storage will provide more flexibility to change and accommodate the research programs. ■

08.27 / LEAN LABS ALLOW FOR SMARTER, MORE EFFICIENT MAINTENANCE.

08.28 / LEAN LABS ARE SAFER LABS.

> "**EDUCATION** probably has more to do with our **ECONOMIC FUTURE** than anything."
>
> —PRESIDENT BARACK OBAMA, SPEECH REGARDING EDUCATION, OCTOBER 2008

09. EDUCATION + PHILANTHROPY

Global Shifts in Education /

Education is the key to solving many of our problems and to creating opportunities. Growth in developing countries will occur only when the population is more educated. **Population growth, better environmental practices, and poverty can be addressed by teaching people, especially at a young age, the latest and best practices.** Studies prove that the more education an individual has, the better chance he or she has for achieving a higher salary and a better quality of life.

My two youngest daughters are being educated in a charter school in Atlanta, Georgia. The school teaches both English and Mandarin, the first school to do so in Georgia. This is good for the children attending the school, but what about the others in the state who do not have the same opportunity? China, a country that spends much less money on education per child than the US, still makes sure that every child learns both Mandarin and English. In most countries, children are taught English as the global language. I predict that by 2020, many more US children will be taught Mandarin in order to compete with other educational programs around the world.

Education, and indeed knowledge, has a monetary value. It can be traded as a commodity and used to purchase other commodities like oil, food, services, or other knowledge. Research universities can be drivers of innovation because they can combine research and education. With strong government support, academic institutions can be the world's leading incubator of innovators. The economic benefits of academic research have been well documented. One of the most comprehensive analyses is by Edwin Mansfield in his paper "Academic Research and Industrial Innovation," in which he concluded that an average annual rate of return to society from academic research ranges from 28 to 40 percent.[1] Universities use their research activities to create new knowledge that supports new products and processes, and educate the next generation of scientists who find jobs in government and private industry.

Data from the Programme for International Student Assessment (PISA) coordinated by the Organization for

Economic Co-operation and Development (OECD) shows how high school students compare to one another internationally in core competencies, and the results are eye opening. Every three years PISA administers a test to 15-year-old students worldwide to measure how well they apply reading, math, and science knowledge in real-world situations. Every period of assessment emphasizes one of the three disciplines, but also tests the other two areas.

The first test, in 2000, focused on reading; the 2003 test stressed mathematics; and the 2006 test emphasized science, covering concepts in physics, chemistry, biology, and earth and space science. The 2006 test was administered to about 400,000 15-year-olds in 57 countries, including the 30 OECD-member countries. These 57 countries make up close to 90 percent of the world economy.

Key findings of the 2006 exam, released in December 2007, are as follows:[2]

- Finnish students, with an average of 563 points on a 1,000-point scale, scored highest on the 2006 science test. It's worth keeping in mind that Finland invests 3.5 percent of its gross domestic product into research and development, the second highest in the world.
- The next highest-performing countries, with mean scores of 530 to 542, were Hong Kong-China, Canada, Chinese Taipei, Estonia, Japan, and New Zealand. Other countries scoring above the OECD average of 500 were Australia, the Netherlands, Korea, Germany, the UK, the Czech Republic, Switzerland, Austria, Belgium, Ireland, Liechtenstein, Slovenia, and Macao-China.
- US students achieved a mean science score of 489, ranking them 21st among the 30 OECD-member countries.
- An average of 1.3 percent of students in OECD countries reached Level 6, the highest proficiency level. These 15-year-olds consistently identified, explained, and applied scientific knowledge in a variety of complex life situations. At least 3.9 percent of New Zealand and Finnish students reached Level 6, three times the OECD average.

The US is losing its competitive edge in education, confirmed the OECD upon releasing the PISA 2006 report. In college graduation rates, "the US slipped between 1995 and 2005 from the 2nd to the 14th rank, not because US college graduation rates declined, but because they rose so much faster in many OECD countries," the report stated. "Graduate output [in the US] is particularly low in science, where the number of people with a college degree per 100,000 employed 25-to-34-year-olds was 1,100, compared with 1,295 on average across OECD countries and more than 2,000 in Australia, Finland, France, and Korea."[3]

> "We go where the
> SMART PEOPLE ARE."
>
> —HOWARD HIGH, INTEL CORPORATION SPOKESMAN

Education is the
GATEWAY TO OPPORTUNITY
and the foundation of a knowledge-based, innovation-driven economy.

—PRESIDENT GEORGE W BUSH,
STATE OF THE UNION ADDRESS, JANUARY 2006

US high-school completion rates also suffer by comparison. "While the US had, well into the 1960s, the highest high school completion rate (76 percent) among OECD countries ... it ranked 21st among the 27 OECD countries with available data," the report added.

In presenting the 2006 report, OECD Secretary-General Angel Gurría said: "In the highly competitive globalized economy of today, quality education is one of the most valuable assets that a society and an individual can have. Skills are key factors for productivity, economic growth, and better living standards. Effective and innovative education policies open enormous opportunities for individuals [and] underpin healthy and vibrant economies."[4] The report goes on to reveal wide variations in skill levels, even given differences in financial resources. "Today, countries like China or India are delivering high skills at moderate cost and at an ever increasing pace," said Gurría.

Although **PISA reports a positive correlation between expenditure per student and performance**, "the relationship is far from straightforward: Finland, New Zealand, Korea, Japan, Australia, and the Netherlands do well with moderate expenditure, while top spenders like the United States and Norway perform below the OECD average," stated Gurría. Across the OECD area, "student performance has generally remained flat between 2000 and 2006, while expenditure on education in OECD countries has risen by 39 percent in real terms during this period."

Socioeconomic background was also found to have significant impact on student performance. **PISA also found that schools with more autonomy, such as those given more responsibility in formulating budgets, tended to perform better. A third major influencer is accountability; schools posting results publicly tend to perform better.** This effect is strong across many countries, suggesting that "external monitoring of standards, rather than relying mostly on schools and teachers to uphold them, can make a real difference to results," said Gurría.

As noted, the USA's decline in education performance is not so much a reflection of what is happening in the USA, but rather it's a result of marked improvements in performance and graduation levels of students in many other developed and emerging countries.

A 2007 National Science Foundation (NSF) special report offers further evidence of the USA's weakened educational stance relative to other nations, especially compared to the impressive emergence of many Asian countries. "The number of people in the world with a post-high-school education has nearly tripled since 1980, with the fastest growth occurring in Asia, including a doubling of both China's and India's percentages," stated the report titled "Asia's Rising Science and Technology Strength."[5]

Given Asia's large—and growing—population, the increased percentage of the continent's college-educated students is highly significant. The increased ratios of first university degrees to college-age populations are documented by the NSF:

- Over the same 12-year period, the EU ratio rose from 11.1 per 100 to 30.7 per 100, bringing it nearly to the US level. The US degree ratio, for decades the highest in the world, rose modestly from 30.9 per 100 in 1990 to 33.9 per 100 in 2002.
- Japan, which has a large higher-education system and a declining population, saw its degree ratio increase from 22.4 per 100 to 32.0 per 100 from 1990 to 2004.
- China's degree ratio stood at 1.2 per 100 in 1990 but jumped to 5.0 per 100 by 2003, on a par with India's 1990 level of 4.8 per 100 in 1990 (no comparable data have been available for India since 1990).
- In South Korea, and especially in Taiwan, degree ratios rose steeply over the decade, equaling or surpassing the ratios of some major industrial nations.

All these numbers do not represent the entire picture. Even though statistics indicate a decrease in many educational categories for the USA there is one important difference—the culture of US education encourages people to continuously challenge existing ideas. Singaporean officials recently visited US schools to learn how to create a system that nurtures and rewards ingenuity, quick thinking, and problem solving. Har Hui Peng stated: "Just by watching, you can see students are more engaged." **The testing ability of students around the world is improving significantly, but their ability to think and collaborate is a more important characteristic for evaluating success.**

When I compare our high schools to what I see when I am traveling abroad, I am terrified for our WORKFORCE OF TOMORROW.

—BILL GATES, CHAIRMAN OF MICROSOFT CORPORATION

Where once nations measured their strength by the size of their armies and arsenals, in the world of the future KNOWLEDGE WILL MATTER MOST.

—PRESIDENT BILL CLINTON

American Education Initiatives /

A 2005 report by the National Academies, the country's leading advisory group on science and technology, sounded an important alarm about the USA's dismal state of science and math education. The congressionally requested report, titled "Rising Above the Gathering Storm," was written by a 20-member committee that included university presidents, CEOs, Nobel Prize winners, and former presidential appointees. The key message was that the erosion of the US's scientific and technical strength is threatening the country's strategic and economic security.

The report argues that as the US falls behind in science education and as it outsources science and engineering jobs, economic crisis becomes inevitable. As of 2005, the ramifications were already clear. A few of many examples cited in the report include the following:

- The USA now imports more high-technology products than it exports. Its trade balance in high-technology manufactured goods shifted from plus $54 billion in 1990 to negative $50 billion in 2001.
- In one recent period, low-wage employers, such as Wal-Mart (now the nation's largest employer) and McDonald's, created 44 percent of the new jobs in the USA while high-wage employers created only 29 percent of the new jobs.
- In 2005, only four US companies ranked among the top 10 corporate recipients of patents granted by the US Patent and Trademark Office.

Unless science and technology capabilities are greatly bolstered, quality of life in the USA will decline rapidly, the report warns. "The United States faces an enormous challenge because of the disparity it faces in labor costs. Science and technology provide the opportunity to overcome that disparity by creating scientists and engineers with the ability to create entire new industries—much as has been done in the past ... But the world is changing rapidly, and our advantages are no longer unique ... **Market forces are already at work moving jobs to countries with less costly, often better-educated, highly motivated workforces and friendlier tax policies.** Without a renewed effort to bolster the foundations of our competitiveness, we can expect to lose our privileged position. For the first time in generations, the nation's children could face poorer prospects than their parents and grandparents did."

In testifying before congress, committee chair Norman Augustine, the retired chair of Lockheed Martin Corporation, said, "It is the unanimous view of our committee that America today faces a serious and intensifying challenge with regard to its future competitiveness and standard of living. Further, we appear to be on a losing path."

The "Gathering Storm" report made four recommendations and identified 20 actions, "including providing federal incentives for promising students to pursue careers in science and math or to teach these subjects in the K–12 system; funding professional development for today's math and science teachers; and increasing federal funding of basic science research by 10 percent each year for the next seven years," according to the National Academy of Sciences.[6] The report also called on the US Department of Energy to sponsor innovative research to meet the nation's long-term energy challenges.

The USA must increase by 10,000 annually the number of capable K–12 science and math teachers, said the committee. Improving the quality of math and science teachers is critical, stated the report, which suggests that more than 60 percent of public school students in some areas of math and science learn from teachers who have not majored in the subject taught or have no certification in it.

"Gathering Storm" is hardly delivering a new message. "A Nation at Risk," issued by a bipartisan federal commission in 1983, warned that the country was engulfed in a "rising tide of mediocrity," citing particularly a "steady decline in science achievement." A steady rain of similar warnings has followed, but this report—on the heels of the Council of Competitiveness' "Innovate America" report in 2004, and the release of Thomas Friedman's *The World Is Flat*, which sent a seismic message about the impact of globalization—has ignited action.

During the 2006 State of the Union Address, President Bush announced the American Competitiveness Initiative (ACI), which aims to strengthen US competitiveness through investment in research and development and education, committed over $136 billion in funding over 10 years. ACI calls for high schools to offer more rigorous coursework in math and science. In promoting best practices in teaching math and science, the administration seeks to train 70,000 high school teachers to lead advanced-placement courses. ACI also seeks to bring 30,000 science and math professionals into the classroom as teachers.

In March 2007 a bipartisan group of senators introduced the "America Creating Opportunities to Meaningfully Promote Excellence in Technology, Education and Science (COMPETES) Act." The bill, which implements recommendations contained in the "Rising Above the Gathering Storm" and the "Innovate America" reports, became law in August 2007, authorizing $43.3 billion over three fiscal years (2008–10) for research and education.

These measures follow President Bush's controversial 2001 No Child Left Behind (NCLB) Act, which mandates that all US students be proficient in math and reading by 2014. Critics say NCLB's regimen of standardized testing encourages schools to "teach to the bottom" at the expense of higher-performing students. "These [gifted] kids don't really count for anything in the federal accountability system," Ann Sheldon, executive director of the Ohio Association of Gifted Children, told US News & World Report in November 2007.[7] NCLB has also been described as "the race to the bottom" as states, in the absence of a single set of national standards, water down their standards to meet federal performance requirements.

Beyond federal actions, other national initiatives are attempting to boost the proficiency levels of US students. Just as the new federal mandates aim to advance the US's economic competitiveness through improved math and science preparedness, the National Governors Association is driving the Innovation America program at the state level with a similar mission. Other powerhouses, such as the Melinda and Bill Gates Foundation, are also heavily committed to improving the outlook for US students. As of 2006, the Gates Foundation, aiming to increase high-school-graduation and college-preparedness rates of low-income and minority students, had funded 1,124 new schools and 761 existing high schools.

At the university level, science educators have the dizzying challenge of keeping up with and teaching

If you **SOLVE THE EDUCATION PROBLEM,** you don't have to do anything else. If you don't solve it, nothing else is going to matter all that much.

—ALAN GREENSPAN, FORMER FEDERAL RESERVE BOARD CHAIRMAN

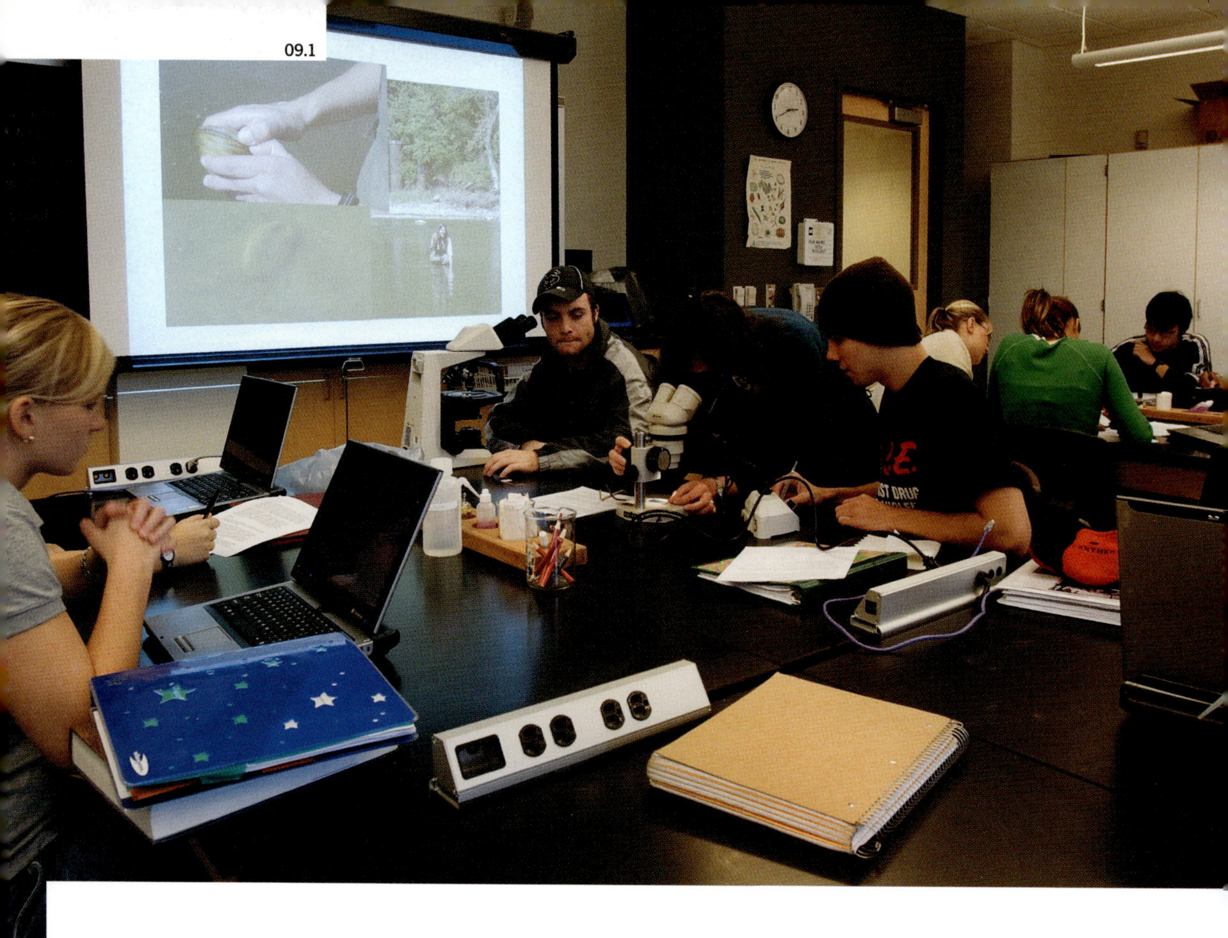

"cutting-edge science," an ever-moving target in this era of "new sciences" and rapid discoveries. The challenge is exhilarating yet daunting. In the following pages we look at a few initiatives that are redefining how science is being taught, and learned, in some of the country's leading educational institutions.

International Universities /

A new model is developing where universities, primarily in the USA and Europe, are expanding overseas collaborating with other universities, private industry, and government agencies. Universities are now competing around the world for the best resources, the most talented teachers, and the brightest students. US universities are well respected around the world. "Overseas programs can help American universities raise their profile, build international relationships, attract top research talent who, in turn, may attract grants and produce patents, and gain access to a new pool of tuition-paying students, just as the number of college-age Americans is about to decline."[8]

It costs much less for students in Qatar to attend a US or UK university in their country than it costs them to go to a university in the USA or the UK. For example, "at Education City in Doha, Qatar's capital, they can study medicine at Weill College of Cornell University, international affairs at Georgetown, computer science and business at Carnegie Mellon, fine arts at Virginia Commonwealth, engineering at Texas A&M, and soon, journalism at Northwestern."[9]

Private industry is also heavily involved to make sure their research centers are next to, and collaborating with, the best academic institutions in the world. "Intel Corporation is adopting this new approach: in 2004, Intel opened lablets—small research facilities—adjacent to three top university research centers, instead of next to its own fabrication facilities. Each lablet is led by a university faculty member who is on academic leave and is not a permanent Intel employee. Intel will not own the output of the research, but hopes to benefit instead by being connected more closely to leading academic research and gaining early access to promising new technologies."[10]

Companies that compete in national and global markets tend to benefit from a virtuous cycle of competition, innovation, and productivity growth. Looking for more creative ways to fund research, both academia and private industry are redefining their partnerships and expanding their long-held approach to intellectual property.

There are concerns faculty will not be as available in their home country because of their support of the international programs. MIT will start a collaborative program with the Singapore government and local university when its new research town, called CREATE, is completed. The project is scheduled to be complete by 2010 with 6 percent of MIT's faculty working in Singapore.

However, a downside to this is that some universities may find it more difficult to maintain their high standards because programs overseas could run the risk of diluting their reputation. Also, will the cross pollination occur within the US university system if students are educated at satellite campuses in their own countries?

Green Chemistry /

Green chemistry and toxicology is increasingly being taught in science courses, requiring students to forge new methods. Academic institutions are also increasing the number of laboratories suited to green chemistry. This generation of students will hopefully lead the world in transitioning to a safer and more sustainable quality of life.

Teaching With Technology /

A good education along with at least basic computer skills prepares people in developed and emerging nations alike for jobs in the 21st century workplace. **Technology is revolutionizing education worldwide, especially in developed countries.** The Internet and burgeoning e-libraries allow everyone to access and to distribute information anywhere. Wireless is everywhere; **distance is becoming irrelevant.** Virtual classes make any wired space on earth a classroom.

The opportunities are infinite and growing exponentially. For example, students and self-learners worldwide have free and open Internet access to Massachusetts Institute of Technology's entire curriculum. MIT's OpenCourseWare project supports the school's mission to advance knowledge globally. Although MIT credit is not earned, anyone may electronically access the syllabi, exams, homework and reading assignments, lecture notes, and, in some cases, lecture videos—materials from about 90 percent of its professors—for all 1,800 MIT courses. The project, begun in 2001, counters the trend for privatization of knowledge and opens the doors to those who cannot afford to pay for a private education.

09.1 / STUDENTS WORKING WITH TRADITIONAL CLASSROOM SPACES WITH NEW TECHNOLOGY AND TEAM-BASED RESEARCH.

09.2 / HANDS-ON PROGRAM: TEAM- AND PROJECT-BASED.

As reported in the *Pittsburgh Post-Gazette*, "You may not have the grades, the money or even the means to get to a physics class with one of the Massachusetts Institute of Technology's

> If you want good manufacturing jobs ...
> GRADUATE MORE ENGINEERS.
> We had more sports exercise majors graduate than electrical engineering grads last year.
>
> —JEFFREY R IMMELT, CHAIRMAN AND CEO OF GENERAL ELECTRIC

best lecturers. But if you have an Internet connection anywhere in the world, you can watch a video of the Dutch-born physics professor, Walter Lewin, swinging on a cable across the front of a lecture hall in his 'Classical Mechanics' course to demonstrate that weight doesn't affect the time it takes a pendulum to complete a cycle of motion."[11]

Open courseware is a hit. By early 2007 the MIT site and sister-sites, that translate the information into other languages, were receiving 1.4 million visits a month from everywhere, including Antarctica and sub-Saharan Africa. The idea of sharing course materials over the Internet is not brand new—the University of California has offered various forms of sharing for years—but the success of the MIT project has given the concept wings. More than 120 universities worldwide are now committed to the open courseware initiative. Among them are John Hopkins, Tufts, Carnegie Mellon, Notre Dame, Utah State, and universities in Spain, China, France, and Japan.[12]

Shared course materials are even popping up on YouTube and iTunes U. A course called Physics for Future Presidents, taught by Berkeley physics professor Richard Muller, "has proven so in demand on YouTube—more than 99,000 people have viewed the first lecture on atoms and heat—that he has a popular book offer from a major publisher," reported the *Pittsburgh Post-Gazette*. "Since the iTunes Store's Web page began on May 30 [2006] listing a couple dozen of the more than 200 universities participating in iTunes U, there have been more than 5 million free downloads."

Open educational resources are growing even faster abroad than in the USA. Korea is "on fire," Vietnam is about to come up with a "very exciting initiative," and China is "very strong and active," John Dehlin, director of the Open Courseware Consortium, a collaboration of more than 100 schools in 20-plus countries, told the *Pittsburgh Post-Gazette* in late 2007.

Visualizations and Virtual Labs /

Another innovative project spearheaded by MIT, in collaboration with Microsoft, is using technology to redefine how education is delivered. The initiative, called iCampus, incubates innovations in educational technology through classroom use and promotes their evolution through worldwide multi-institutional cooperation. Begun in 1999, the seven-year iCampus partnership resulted in several successful and ongoing projects. One project created a studio-based physics classroom that facilitates hands-on experimentation. Another project makes specialized laboratory equipment available worldwide through the Internet.

iCampus sought to realize technology's real potential in education. "Education looks like the Industrial Revolution, where you have classrooms designed around some kind of factory model," stated Randy Hinrichs, a manager at Microsoft Research, on research.microsoft.com.[13] "Everyone turns to page 37 at the same time." Instead of using computers to their full capabilities—to deconstruct objects, build a bridge, look inside of a molecule, and recreate the human body digitally—classrooms largely use them as electronic books, said Hinrichs.

Noting the popularity of computer games, Microsoft and MIT developed Games to Teach to help students understand basic principles of physics. "They just didn't see it, they couldn't experiment with things," said Hinrichs. An MIT professor designed simulations to illustrate principles, such as Faraday's Law, allowing students to "actually see the forces that were imposed on the experiments," reported Microsoft Research.

Another iCampus-funded project, Technology-Enabled Active Learning (TEAL), combines a studio format, hands-on lab experiments, visualizations, and small-group collaboration into the classroom experience. Students have access to computer simulations as well as to networked electronic whiteboards that allow them work in groups, save their whiteboard scribbles, and get feedback from professors locally or remotely in real-time.[14] Launched in 2000, TEAL has produced consistent and measurable learning gains, reports MIT.

iCampus realizes a different paradigm of teaching. Instead of just a lecture format, a mechanical engineering course now presents "a 15-minute talk followed by interactive hands-on experiments, followed by a component that's online and simulated," iCampus program manager Dave Mitchell reported to Microsoft.[15]

Another MIT initiative, online laboratories (iLabs), puts state-of-the-art instrumentation online so that students worldwide can access sophisticated machines not available at their local colleges. Students can get real-time data to incorporate into their own projects. The iLabs vision is to share expensive equipment and educational materials associated with lab experiments as broadly as possible. The ultimate goal is to establish a worldwide economy of shareable labs to enhance science and engineering education in a broad range of disciplines.

iLabs eliminate many logistical and economical limitations of conventional laboratories. Unlike conventional labs, which must be owned and maintained by each institution, iLabs can be shared worldwide around the clock, allowing many students to be supported on limited equipment. Furthermore, students need not be present in potentially unsafe experimental facilities.

One of MIT's first iLabs was a Microelectronics Device Characterization test station that enables students to take measurements of the current/voltage characteristics of transistors and other microelectronic devices.[16] This iLab is now used in three different courses by more than 500 students. In all, more than seven new iLabs were created, including those for use in microelectronics, chemical engineering, polymer crystallization, structural engineering, and signal processing. In 2005 these iLabs were used for credit-bearing assignments by students in China, Egypt, Greece, Italy, Nigeria, Sweden, Taiwan, Tanzania, Uganda, and the UK. To date, more than 5,500 students have used iLabs, and schools worldwide are using MIT's iLab Shared Architecture software framework to develop new iLabs.

Insidehighered.com describes an iLab experience like this: "A bleary eyed Massachusetts Institute of Technology student roaming the ground floor of Building 1 in the wee hours might hear a rumbling coming from the 'shake table' lab, where researchers simulate earthquakes. But a peek into the lab would reveal that nobody is there. More than 5,000 miles away at Obafemi Awolowo University, in Nigeria, morning classes have begun, and a student might have just loaded the Kobe Earthquake simulation, turned on the lights in the Cambridge lab, and is gathering data from MIT's equipment."[17]

"Working internationally teaches you how to communicate with others even with a big difference in culture," MIT student Scot Frank stated in an MIT release.[18] "There are different teaching methodologies between the two countries but we really learn from each other. It's really collaboration."

As Kevin Schofield, general manager of Microsoft Research Strategy and Communications reported to research.microsoft.com, "One of the things I've learned from this is that educational resources, particularly the most valuable ones, are scarce and unevenly distributed around the world. That includes not just lab equipment and computers and networks, but, probably most importantly, faculty. The ones who are really experts on any one particular topic are geographically dispersed but not evenly." In parts of India and China, for example, "there just aren't enough teachers, and there never will be," continued Schofield.[19] Projects such as iLabs can make resources and faculty globally available.

Shanghai Jiao Tong University, one of two international ranking systems, ranked the Institutes of Higher Education for 2007. The top 109 universities are found in:

Country	Universities
USA	53
EUROPE	43
ASIA	7
CANADA	4
AUSTRALIA	2

The top 10 are listed with USA holding eight of 10 spots, and Europe the remaining two. Tokyo University ranked the highest in Asia at number 20. For China, National Taiwan University ranked the highest at 163. Mainland China's highest ranking is Tsinghua University at 169. India has very poor representation with its highest ranking for the Indian Institute of Science at number 316.

There is also a European ranking of universities worldwide. The following 2007 list shows the UK with a stronger presence in the top 10 but the overall top 100 is similar to the ranking by Shanghai Jiao Tong University.

Rank	Institution	Country
1.	HARVARD UNIVERSITY	USA
2.	UNIVERSITY OF CAMBRIDGE	UK
2.	UNIVERSITY OF OXFORD	UK
2.	YALE UNIVERSITY	USA
5.	IMPERIAL COLLEGE	UK
6.	PRINCETON UNIVERSITY	USA
7.	CALIFORNIA INSTITUTE OF TECHNOLOGY	USA
7.	UNIVERSITY OF CHICAGO	USA
9.	UNIVERSITY COLLEGE LONDON	UK
10.	MASSACHUSETTS INSTITUTE OF TECHNOLOGY	USA

We now live in a world that is becoming more competitive each year. Major decisions made by politicians, as well as those made by leaders in academia and private industry, will become even more critical and will have a greater impact on all of us, whether it be in a positive or a negative direction. Over the next decade academic institutions around the world will need to meet new challenges by becoming more focused and developing their core strengths in order to meet these demands.

Global Educational Outreach /

Global Educational Outreach for Science, Engineering and Technology (GEOSET) has been developed by Nobel Prize winning scientist Harry Kroto, faculty at Florida State University as well as Sheffield University in the UK.

GEOSET is another revolutionary approach to educational outreach. Teachers and students all over the world can have instant access, free of charge, to a permanent Internet library of outstanding educational material. With a primary focus on science, engineering, and technology, the aim of the GEOSET project is to create a resource that will revolutionize education—particularly science education—on a global scale. A network of collaborating universities and other educational establishments will make a wide range of directly downloadable educational material available, along with audio-visual guides on how to best use the material in the classroom. Each participating institution can focus on its own creative concepts; however, by working together all the bases can be covered.

The target audience is primarily the large number of dedicated teachers struggling with minimal resources and limited training. However, the approach also serves an exceptionally wide range of other educational outreach and communication needs. The pilot Global Educational Outreach (GEO) initiatives, already in place at Florida State University, USA, and Sheffield University, UK, are proving to be highly effective, especially for science, engineering, and technology (SET).

The advantages of using GEOSET are significant:

- The new technology allows teachers to exercise their creativity, thereby fostering creativity in their students.
- Preparation time is significantly reduced for teachers, who will have instant access to modular STEM teaching materials that are designed either to be used directly in the classroom or as a high-quality presentation.
- GEOSET offers teachers expertly produced material in areas that may lie outside their specialties. (For example, a committed biology teacher may be the best person available in a school to teach physics even if he or she doesn't have a specialized background in the subject.)

The GEOSET project has been developed on the basis of considerable experience in creating over 150 television and Internet science programs with the Vega Science trust (www.vega.org.uk), more than half of which have been broadcast on the UK's BBC. Formal and informal agreements to participate in GEOSET have already been made with institutions in Canada, Europe, India, Japan, and the USA.

GEOSET offers presenters a great degree of creative freedom to produce educational material with relative ease, and for the users the material is instantly available, permanent, and free of charge. In the same way that the invention of the printing press and free access to books fostered the intellectual creativity that made the Enlightenment possible, the Internet has democratized broadcasting. Using this freedom, GEOSET makes downloadable teaching material readily available from reliable sources of knowledge.

One of the goals is to assemble a body of material that will produce a network of collaborating, interactive GEOSET sites worldwide. GEOSET will help to bridge divisions between science, the humanities, and the arts, as well as the ones that exist between cultures. As the scheme evolves, the repertoire of Internet technologies will expand to include processes such as video on demand (VOD), video web conferencing, and audio and video podcasts.

Partners in Learning /

Education is a cornerstone of economic and social opportunity, yet education systems around the world still face serious challenges. Many countries are struggling just to provide the infrastructure and fundamentals needed to meet existing or growing demands such as classrooms, qualified teachers, curricula, and technology. At the same time, countries are being challenged to modernize and tap into the global economy in order to be competitive.

Partners in Learning (PiL), sponsored by Microsoft since 2003, is a US$250-million, five-year global initiative designed to increase technology access for schools, foster innovative approaches to pedagogy and teacher development, and provide education leaders with the tools to envision, implement, and manage change. The program has touched the lives of more than 80 million students, teachers, and education policymakers in 101 countries.

PiL is based on collaborative agreements with governments and non-governmental organizations. The key elements of the initiative are:

- technology access
- curriculum and training resources
- grants to governments and educational organizations
- forging new models for teaching and learning.

One thing the programs have in common is the potential to spark systemic change, particularly in more traditional education systems. Programs are being implemented everywhere around the world from cities to remote villages.

One PiL program in the UK is Enquiring Minds, an innovative approach to learning that looks at different goals beyond test results and enables children to become effective researchers, innovators, and creators of knowledge. It incorporates digital tools such as collaborative software, digital cameras, and laptop computers. The results from these pilot schools will inform a broader transformation of curricula in the UK that is part of Building Schools for the Future, the government's effort to rebuild or renovate every secondary school in the country in the next 10 years.

In Singapore, digital multimedia textbooks have replaced some of the hardcover versions. This technology is supported by researchers and developers who monitor the Tablet PCs while students use them in the classroom, and from the information they gather are able to develop new applications that further enhance teaching and learning. The hands-on research has encouraged educators to develop and test advanced technologies in real classroom environments.

To recognize Argentina's most innovative teachers, the team produced a series of more than 100 television documentaries that shared those teachers' best practices with more than 250,000 educators around the country. In 2007 the team also rolled out the Peer coaching program nationwide that reached 4,000 teachers in an effort to promote ongoing collaboration and professional development.

In 2008 alone, up to 60,000 new teachers in India received ICT skills training through Microsoft's pre-service initiatives. In turn these teachers reach an estimated 30 million students. One of the objectives of this program is to bridge the enormous gap between rich and poor.

There are also several success stories in the USA. Project-based teaching methodologies, innovative uses of technology, and a focus on equipping students with lifelong learning skills are hallmarks of the progressive educational environment taking shape at the Philadelphia School of the Future (a state-of-the-art public high school). The 750-student school was built to maximize collaboration, foster creativity, and teach real-world skills. Students and teachers have extensive access to information technology and other digital tools that help support creative teaching methods and continuous, self-directed learning. Recognizing that students absorb knowledge in different ways, the school supports adaptive instructional methods geared toward individual learning needs.

PiL has taken a broad approach, supporting efforts to integrate technology effectively into teaching and learning, and programs that help school leaders and education policymakers create an environment in which school systems can be more agile, responsive, and efficient.

PeaPoD: Portable, Educationally Adaptive, Product of Design /

The World Bank estimates that educating children worldwide will require the construction of 10 million new classrooms in more than 100 countries by 2015. That is a rate of 2 million new classrooms per year for the next five years.

The 2009 Open Architecture Challenge, sponsored by Architecture For Humanity, invited the global design and construction community to collaborate with primary and secondary school teachers and students to design classrooms that try to help tackle this challenge.

One of the winning competition entries was a classroom called the **PeaPoD, which stands for Portable, Educationally Adaptive, Product of Design. The concept of delivering portable classrooms anywhere in the world** is one of the ways to fulfill the tremendous need for classrooms, especially in developing countries. They are very quick to install and do not depend on scarce materials or special skills to construct. Conventional portable classrooms are widely used throughout the United States because they are inexpensive relative to conventional construction and provide short-term solutions to what are often times long-term problems. However, most portable classrooms do not adequately address the needs of the students nor address issues of global sustainability.

The PeaPoD relocatable classroom is designed specifically to enhance learning by creating a classroom environment that can be adapted to various teaching styles and environmental conditions. **The PeaPoD classroom is mass produced in a factory and can be shipped anywhere in the world. The idea that a student in a developing country could learn in the same classroom environment and receive the same benefits as a student in the United States is a powerful one.** By using passive techniques such as adequate daylight, operable windows, a water catchment, and natural ventilation, the PeaPoD can operate with little or no utility costs while at the same time providing a learning environment that increases test scores. Making

09.4

sure these classrooms are sustainable in the way they are designed, built, and operate is critical. **Studies have shown that students who learn in green schools and classrooms show a 20 percent improvement over students tested in non-green schools.** Just bringing natural daylight into the classroom helps students improve on test scores. In one study, the classroom with the most daylight progressed 20 percent faster in one year on math tests and 26 percent faster on reading tests than those students who learned in environments that received the least amount of natural light. Another extremely important factor in providing proper learning conditions around the world is a healthy learning environment. Good indoor air quality is achieved by specifying renewable materials that do not contain harmful volatile organic compounds (VOCs) and by allowing adequate fresh air into the classrooms. **Similar to daylight in a classroom, proper indoor air quality has a dramatic impact on students' performance in the classroom. The PeaPoD helps to address the need for a safe and healthy learning environment on a global scale.**

09.3 / PEAPOD ENTRY SHOWING BUTTERFLY ROOF, WATER CISTERNS, CLASSROOM GARDEN PATIO, AND OUTDOOR WRITABLE WALL.

09.4 / INTERIOR VIEW SHOWING ADAPTABLE FURNITURE LAYOUT, BREAKOUT AREA, AND OPERABLE WINDOWS.

09.5

Biologically Inspired Design: A Tool for Interdisciplinary Education /

In the following section are thoughts I have put together from conversations with Jeannette Yen, who is the Director at the Center for Biological-Inspired Design (CBID) at Georgia Institute of Technology.

As the world becomes increasingly interdisciplinary, it has become clear that we do not fully understand how to foster, sustain, and propagate these interdisciplinary connections in the university. Such interdisciplinary education is necessary to solve the increasingly complex problems faced by **the next generation of scientists and engineers, who in particular, will be faced with the challenge of integrating biology and engineering. Biologically inspired design (BID) is a catalyst for innovation, where engineers embrace biology as a source of inspiration and biologists utilize the design space of engineers.** Innovations occur in diverse fields such as biomaterials, bio-inspired locomotion, bio-sensors, systems organization, and green technology. This approach relies heavily on experiential problem-based learning to design prototypes of biomimetic processes to test and construct, facilitated by a forum of experts (such as is available in the faculty of our Center for Biologically Inspired Design at Georgia Institute of Technology). These devices will test biological functions and provide new ideas for engineering solutions. This approach requires engineers and biologists to develop a common language and mode of thinking that facilitates collaborative problem identification and solving. Professors from biology, polymer engineering, industrial ecology, mechanical, and materials science engineering: practitioners of BID serve as models of collaboratories, where people work together to examine evolutionary adaptation as a source for design inspiration, utilizing principles of scaling, adaptability, and robust multifunctionality that characterize biological systems. In an educational program, the outcome includes: a model for practicing interdisciplinary exchange; an assessment of what nature can offer to engineering; an evaluation of BID as a catalyst for innovation; the development of the language of interdisciplinary research. The use of BID as part

09.5 / THE SPIDER WEB IS MADE OF SILK MATERIAL. IT IS STRONG, CONTINUOUS AND INSOLUBLE IN WATER. THE SPIDER WEB IS AMAZINGLY RESISTANT TO RAIN, WIND AND SUNLIGHT. THE SPIDER WEB SILK IS APPROXIMATELY THREE TIMES STRONGER THAN STEEL. RESEARCHERS TODAY ARE TRYING TO CREATE FIBERS, LIKE SPIDER WEBS, THAT WILL BE COMMERCIALLY VIABLE.

of the toolkit in engineering design courses can result in novel solutions thus propagating the utility of this tool. The desired output is: the 21st-century Engineer, an inventor equipped with core strengths in engineering, yet also able to speak the language of biology and learn from its structures, processes, and principles; and the 21st-century Biologist, a naturalist able to investigate function using engineering tools, team with engineers and physical scientists to examine biological structures, processes, and systems in terms of evolved solutions to problems, and as sources of engineering designs.

Training students with these skills will facilitate the invention of novel processes or products that increase the global competitiveness of our industries while increasing our energy independence, decreasing our reliance of toxic materials, and reducing our ecological footprint. The Center activities provide a source of biologically inspired technology and technical education that can be linked to a larger global network of institutions and industries that practice biologically inspired design. Such a center can serve as a magnet, and recruiting tool for the best interdisciplinary students and faculty in the sciences, engineering, and design fields, and provide significant opportunities for increasing the participation of underrepresented groups.

There are many studies that are beneficial to get people to look and think differently and hopefully more effectively. For example, any child is familiar with the spider's web. The beauty of the dew outlining the pattern spun of silk makes us so aware of this living creature. We can take this awareness and tweak it by saying something like, did you know that silk is as strong as steel? Now a curious kid will show a puzzled look and then sneak off to pull a web apart and return with a challenge, saying it's not so. To answer, we can take light weights to pull on a web and show how strong it is, thus introducing the quantitative analysis of silk strength. Then with that measure of strength, we can introduce the concept of scale: if you had as much silk as is in a steel cable, how strong would that silk cable be? Et voila! You'll have shown the skeptic the strength of this natural substance made a room temperature of protein manufactured by a living organism, the spider. All created without toxic wastes and built on local supplies.

Other comparable educational opportunities exist at:

- Wyss Institute for Biologically Inspired Engineering, Harvard
- Stanford BioX program
- UC Berkey's CIBER PROGRAM and IGERT
- Duke Center for Biologically inspired materials

Philanthropy /

Over the past quarter century, largely due to the technological revolution, many Americans have become extremely wealthy. As of 2006, about 7,000 US households had amassed more than $100 million in wealth; about 500,000 had $10 million or more.[20] Wealthy philanthropists and private foundations are funding science on a grand scale.

Private Foundations /

Private foundation funding for biomedical research increased 36 percent between 1994 and 2003 to around $2.5 billion, according to a study published in 2005 in *The Journal of the American Medical Association*. The author of that paper, Hamilton Moses, later estimated private foundation funding in all its varieties to be over $5 billion.[21]

It's no wonder that some of the wealthiest Americans have selected science-related endeavors as their pet causes. Science, after all, is what made many of them rich. Among the top five donors in US history are Bill Gates of Microsoft, William Hewlett and David Packard of Hewlett-Packard, and Gordon Moore of Intel.[22] All of their foundations contribute heavily to science.

Other top-10 ranking US philanthropists are major science backers as well, even though their fortunes were not built strictly on science and technology: Warren Buffett of Berkshire Hathaway gifted the Bill and Melinda Gates Foundation a whopping $30 billion; Herbert and Marion Sandler of Golden West Financial are, through their $1.3-billion foundation, one of the nation's largest supporters of medical research and environmental causes; and Jim and Virginia Stowers of American Century have contributed over $1 billion to the Stowers Institute of Medical Research.

Science philanthropy in the USA is not new, of course. It was jumpstarted at the dawn of the 20th century with the creation of the Carnegie Institution in 1902 and the Rockefeller Foundation in 1913.

Andrew Carnegie caused a sensation when he announced in 1901 that he would fund scientific discovery to the tune of $10 million—much more money than even the US government spent on basic science at that time. It bothered Carnegie that the USA was undistinguished in science and that US universities lacked the solid research tradition common to many European institutions.

The institute, now valued at more than $635 million, studies a broad range of topics, including embryology, developmental molecular biology, plant biology, ecology, cosmology and astrophysics, geochemistry, cosmo-chemistry, and crystallography. Among its many famed researchers are: Edwin Hubble, who revolutionized astronomy with his discovery that the universe is expanding and that there are galaxies other than the Milky Way; Charles Richter, who created the earthquake measurement scale; and Vera Rubin, who was awarded the Presidential Medal of Science for her work confirming the existence of dark matter in the universe.

The Rockefeller Foundation also supported an all-star cast, including Hermann Muller, Max Delbrück, T. Caspersson, E. Hammarsten, Linus Pauling, Robert Corey, T. Svedberg and W.T. Astbury. By mid-century, these scientists had transformed biochemistry and genetics, setting the stage for the Watson–Crick double helix to emerge in 1953. The birth of modern biology was truly midwifed by the Rockefeller Foundation.[23]

From the 1890s to the 1940s private foundations were the major independent source of funding for all biomedical research. In 1940, foundations provided 27 percent of the $45 million spent on health-related research.[24] After the Second World War, federal agencies assumed the dominant role in funding scientific research, although philanthropic funding remained important.

The Howard Hughes Medical Institute (HHMI), chartered in 1953, is one of the largest private medical research organizations in the world. In the USA, only the federal government spends more money for medical research than the nearly $500 million received annually by some 330 HHMI scientists. In 2007, its endowment exceeded $16 billion. The institute's scientists conduct research in six broad areas: genetics, immunology, cell biology, neuroscience, structural biology, and computational biology. A few of the important discoveries made by its scientists include: the genes responsible for cystic fibrosis and muscular dystrophy; a non-invasive test for colon cancer; a drug that fights leukemia; breakthroughs in AIDS research; and work that may lead to a cure for spinal-cord injuries.

The USA's earlier magnates didn't spawn foundations as large as the Gates Foundation, even after adjusting for inflation, yet their efforts have had a lasting impact. Coupled with today's bounty of gifts to science, wealthy foundations are playing an unprecedented role in advancing science. The money is coveted by research institutions, which have been parched by flat funding from the US National Institutes of Health and other sources in recent years.

The Bill and Melinda Gates Foundation, which has grown considerably with the aid of Warren Buffett, is the world's richest and arguably the most dynamic charity. Guided by the belief that every life has equal value, the Gates Foundation works to improve health, reduce extreme poverty, and increase access to technology worldwide.

Gates Foundation Investments (as of 2009):

GLOBAL HEALTH PROGRAMS	$13,049,725,067
GLOBAL DEVELOPMENT	$2,823,713,536
UNITED STATES	$5,798,073,920
CHARITABLE SECTOR SUPPORT	$22,526,050
EMPLOYEE MATCHING GIFTS & SPONSORSHIPS	$14,916,040
FAMILY INTEREST GRANTS	$897,982,479

The foundation's global health mission is to ensure that lifesaving advances in health are created and shared with those who need them most. The foundation works to accelerate access to vaccines, drugs, and other tools to fight diseases that disproportionately affect developing countries. It also supports research to discover effective, affordable, and practical health solutions for people in poor countries.

Product Development Partnerships /

A product development partnership (PDP) is a non-profit organization that builds partnerships between the public, private, academic, and philanthropic sectors to drive the development of new products for underserved markets. Through their unique, collaborative efforts, PDPs are able to access a variety of funding sources, and to apply a wide range of tools and knowledge to their programs. PDPs retain direct management oversight of their projects, though much of the work is done through external research facilities and contractors. In the global health arena, PDPs have been established to accelerate the development of new technologies that fight TB, AIDS, malaria, and a wide range of neglected diseases. PDPs are created for the public good; their products are made affordable to all those who need them.

The TB Alliance is a PDP that combines the research and development expertise of its own staff with the skills and resources of its partners to streamline and accelerate TB drug development. The organization manages a portfolio of candidate TB compounds, from both public and private sector sources, using a variety of licensing and partnership agreements. Such cooperative efforts allow TB Alliance to apply the latest scientific advances, and the highest technical capabilities, to the drug development efforts. By operating as a virtual research and development organization to minimize costs, including overhead and investments in infrastructure, the business model helps to optimize the speed of development.

Less than 3 percent of global funding for health research and development is dedicated to diseases of the developing world. Over the past decade, global health PDPs like the TB Alliance have advanced dozens of potential new diagnostics, drugs, vaccines, and microbicides through the development pipeline, toward registration and launch. PDPs are able to mobilize private sector participation to develop new technologies where there is no market incentive. PDPs also fill a crucial gap—basic academic research in non-profitable fields is translated into actual product development, evaluation, and use with the potential to save millions of lives.

PDP Donor Funding

BILL AND MELINDA GATES FOUNDATION	50%
US AGENCY FOR INTERNATIONAL DEVELOPMENT	18%
UK DEPARTMENT FOR INTERNATIONAL DEVELOPMENT	14%
NETHERLANDS MINISTRY OF FOREIGN AFFAIRS	7%
ROCKEFELLER FOUNDATION	6%
IRISH AID	5%

Other Projects /

Fred Kavli is a retired technology entrepreneur who aims to spark a renaissance in basic research in nanoscience, astrophysics, and neuroscience. With his $600 million coffer, he has launched 14 of 20 planned research centers in top universities, including Harvard, Yale, Caltech, and MIT, plus schools in Europe and China.[25] He has also created what he hopes will be the 21st century's version of the Nobel Prizes. The Kavli Prizes, which began in 2008 and are presented every two years, award a total of $3 million, or $1 million to scientists in each of the three fields of nanoscience, astrophysics, and neuroscience.

Kavli's fascination with science began in his youth on a farm in Norway where he grew up surrounded by nature and was spellbound by the Northern Lights. After earning a physics degree in 1955, Kavli emigrated to California.

"I used to ski across the vast white expanses of a quiet and lonely mountaintop," Kavli said when he announced his prizes in 2006 in Oslo. "At times, the heavens would be aflame with the Northern Lights, shifting and dancing across the sky and down to the white-clad peaks. In the stillness and solitude ... I pondered the mysteries of the universe, the planet, nature and of man. I'm still pondering."

Another philanthropist impassioned by physics was William I. Fine (1928–2002), a developer who made possible the Fine Theoretical Physics Institute at the University of Minnesota. More recently, Mike Lazaridis, founder and co-CEO of Research In Motion (of BlackBerry fame), donated US$94 million to create the Perimeter Institute for Theoretical Physics in Waterloo, Canada.

Many people believe that, dollar for dollar, donor money is more valuable than government or industry funding. **Private foundations have more flexibility in spending and their backers are apt to be big thinkers and risk takers.** These new "gigaphilanthropists" have a disproportionate influence. "They can, and do, take financial and scientific risks unthinkable with tax-payers' dollars," Hamilton Moses of the Alerion Institute, a Virginia-based biomedical research think tank, told *Nature* magazine. [26]

Private givers are largely not answerable to shareholders, venture capitalists, politicians, or the public at large. They are able to support controversial or unpopular research. Retired Silicon Valley venture-capitalist Andrew Rachleff is a prime example. In January 2008, he and his wife, Debra, endowed a fund plowing $9 million into high-risk cancer-research projects that had no chance of government or industry funding. "You don't get great reward without taking great risks," Rachleff told the *San Francisco Chronicle*.[27] The culture of the National Institutes of Health is "old guys giving money to old guys," he continued. Instead, funding should be directed to the young scientists who are "swinging for the fences." To beat cancer and other diseases, **cautious grant-making agencies must think like venture capitalists, Rachleff said.**

Computer industry philanthropists often see the progress of disease research as excruciatingly slow. Intel co-founder Andy Grove certainly does. Diagnosed with Parkinson's disease, he has pledged $40 million to the Michael J. Fox Foundation for Parkinson's Research.

Going forward, we can only expect such donations to grow. Before 2040, the USA will experience the largest intergenerational transfer of wealth in history. A conservative estimate of the amount of money due to be inherited by baby boomers from their frugal, postwar parents is $10 trillion, and about 10 percent is expected to be set aside for philanthropic purposes.[28]

Funding, and more of it, is critical. Advances in science speak directly to the greatest challenges facing the world, as eloquently stated by Susan M. Fitzpatrick and John T. Bruer in a 1997 *Science* magazine editorial: "As we worry about the needs of an aging population, neuroscience and genetics are unraveling the mysteries of Alzheimer's disease. As we struggle with how to provide quality education to increasingly diverse populations, scientists are learning much about how the human mind develops and learns. As we deal with increasingly global problems, researchers are developing sophisticated tools to model complex systems such as climate, population, and the emergence of disease that can be used to guide policy and inform decision-making. At the same time, fiscal constraints make it unlikely that the federal government can significantly increase its support of science."[29] We must therefore look to private foundations to support science in the spirit of Carnegie, Rockefeller, and the new breed of leading philanthropists.

10. IDEAS

Global Mission and International Collaboration /

For research to truly become a global endeavor, all nations should strategically join in the development of the new sciences. And within those nations, governments, business concerns, philanthropists, and academia should focus on how to improve the "science of science." Such global collaboration will accelerate discoveries and speed their benefits for the greatest good. The opportunities to improve quality of life, length of life, and world economics are vast.

Four ways to foster international collaboration, each of which is discussed below, are as follows:

1. Create a global research partnership
2. Enhance research and development incentives
3. Commit globally to environmental research and sustainability
4. Develop and connect research facilities worldwide.

Create a Global Research Partnership /

As authors Don Tapscott and Anthony D. Williams so aptly exemplified in their 2006 bestseller *Wikinomics: How Mass Collaboration Changes Everything*, collective participation is a mighty force. Innovative global collaboration was at play when Linus Torvalds encouraged thousands of anonymous programmers to critique computer code and add value on their own. The result was Linux software for personal computers.

That same spirit launched The Scripps Research Institute Olson Laboratory, the first-ever biomedical distributed computing project. The Fight AIDS at Home project enables any Internet-connected person to use free, downloadable biomedical software to help scientists develop drugs to control HIV and advance the goal of preventing AIDS. The software uses a computer's idle cycles to simulate the attachment process of molecules to the HIV-1 protease. Scientists then study any promising findings.

Mass collaboration also fueled Proctor & Gamble's successful InnoCentive program offering freelance scientists cash for solutions. "**Just register on the InnoCentive network where**

you and 90,000 other scientists around the world can help solve tough R&D problems for a cash reward," wrote Tapscott and Williams in *Wikinomics*. "With the right approach, companies can obtain higher rates of growth and innovation by learning how to engage and co-create with a dynamic and increasingly global network of peers."[1]

The power of mass collaboration can be put to work in the formation of a global research partnership. Similar to the World Health Organization (WHO), worldwide leaders in politics, private industry, and academia should develop an international strategy to support important research, with an emphasis on efficiency, effectiveness, affordability, and collaboration. Such a coalition should provide incentives, value-added partnering support, and excellent databases and communication tools to spur advancements and share successes.

Business models exist for effective global partnerships. For example, Sanofi-Aventis, Europe's largest pharmaceutical company, is a world leader in creating partnerships to provide affordable medicines to the developing world. In 2004, the multinational launched Impact Malaria, a program to develop and provide drugs in partnership with non-government organizations in affected regions. In 2006, Sanofi committed US$25 million to sponsor WHO's work treating neglected tropical diseases, such as sleeping sickness. Other Aventis partners include the Global Alliance for Vaccines and Immunization, the Global Polio Eradication Initiative, and scores of health-related companies. Aventis's products are increasingly essential in battling epidemics worldwide, but partners such as WHO, with unique capabilities and access, are needed to distribute medicines and monitor progress. Clearly, such partnerships add value and expand possibilities.

10.1 / FARMER MAMO TESFAYE HAS "LEOPARD SKIN" ON HIS LEGS FROM DEBILITATING ONCHOCERCIASIS, OR RIVER BLINDNESS DISEASE. THANKS TO ANNUAL TREATMENTS OF MECTIZAN®, DONATED BY MERCK AND CO, INC., MAMO IS ONCE AGAIN ABLE TO FARM. PHOTOS COURTESY OF THE CARTER CENTER AND THE ETHIOPIA RIVER BLINDNESS PROGRAM.

10.2 / NORTH SHOA, ETHIOPIA. DISTRIBUTION OF LONG LASTING INSECTICIDE NETS (LLIN) TO PREVENT MALARIA, THE COUNTRY'S BIGGEST KILLER DISEASE OFTEN ARRIVE IN BALES VIA CAMEL BACK.

Enhance Research and Development Incentives /

Globally, less than 10 percent of money invested in pharmaceutical research and development is targeted to diseases of poverty, such as malaria, AIDS, and tuberculosis. Yet, these diseases affect up to 90 percent of the world, according to the Global Alliance for TB Drug Development.[2] To make a profit, pharmaceutical companies instead focus on "lifestyle" drugs and treatments for "Western" diseases.

To develop low-cost medicine for the world's poor, biotech and pharmaceutical companies must be profitable. Enhanced tax incentives are needed to entice these companies to accelerate discoveries and develop affordable vaccines and other medicines.

"We've got to make some deal with [pharmaceutical companies] that allows them to do what they do well and still gets this medicine to people in the [shortest] possible time at prices they can afford," former President Bill Clinton said while addressing the 2005 Time Global Health Summit.[3] Companies can also reduce risk and accelerate discoveries by pooling intellectual property and sharing in financial gains.

Private industry plays a vital role in addressing world health needs. By 2006, in its first six years, the Global Alliance for Vaccines and Immunization had immunized some 138 million children and prevented more than 2.3 million premature deaths, reported WHO. The vaccines and immunizations were primarily developed and donated by biotech and pharmaceutical companies.

10.3

Country	R&D Tax Incentive	Comments
Australia	ALLOWS A 125 PERCENT DEDUCTION (EQUAL TO A FLAT 7.5 PERCENT TAX CREDIT) FOR R&D EXPENDITURES, PLUS A 175 PERCENT DEDUCTION FOR R&D EXPENDITURES EXCEEDING A BASE AMOUNT OF PRIOR-YEAR SPENDING.	DUE TO ITS R&D-FRIENDLY ENVIRONMENT, 50 PERCENT OF THE MOST INNOVATIVE COMPANIES IN AUSTRALIA ARE FOREIGN-BASED, THE GOVERNMENT REPORTS.
Canada	OFFERS A 20 PERCENT FLAT R&D TAX CREDIT. MANY PROVINCIAL GOVERNMENTS ALSO OFFER INCENTIVES, SUCH AS REFUNDABLE CREDITS.	IN 2003, US SUBSIDIARIES SPENT US$2.5 BILLION ON R&D IN CANADA, WHICH PROMOTES "R&D TAX CREDITS, AMONG THE MOST GENEROUS IN THE INDUSTRIALIZED WORLD" AND A LOW COST STRUCTURE.
China	FOREIGN ENTERPRISES RECEIVE A 150 PERCENT DEDUCTION IF R&D SPENDING IS UP BY 10 PERCENT FROM THE PRIOR YEAR.	THE 10 PERCENT INCREMENTAL-INCREASE THRESHOLD IS NOT DIFFICULT FOR START-UP OPERATIONS TO MEET.
France	A 50 PERCENT R&D CREDIT INCLUDES A 5 PERCENT FLAT CREDIT AND A 45 PERCENT CREDIT FOR R&D EXPENDITURES IN EXCESS OF AVERAGE R&D SPENDING OVER THE TWO PREVIOUS YEARS.	FRANCE IS REAPING THE BENEFITS OF A NEW TREND—AMERICANS GOING TO EUROPE FOR R&D ACTIVITIES. IN 2003, US SUBSIDIARIES SPENT US$1.8 BILLION ON R&D IN FRANCE. THE INCREMENTAL THRESHOLD IS EASY FOR GROWING COMPANIES TO MEET.
India	SCIENTIFIC R&D COMPANIES ARE ENTITLED TO A 100 PERCENT DEDUCTION OF PROFITS FOR 10 YEARS. THE AUTO INDUSTRY RECEIVES A 150 PERCENT DEDUCTION FOR IN-HOUSE R&D EXPENDITURES.	MORE THAN 100 GLOBAL COMPANIES ESTABLISHED R&D CENTERS IN INDIA IN THE FIVE YEARS PRECEDING 2005. "IF INDIA PLAYS ITS CARDS RIGHT, IT CAN BECOME BY 2020 THE WORLD'S NUMBER-ONE KNOWLEDGE PRODUCTION CENTER," RAGHUNATH MASHELKAR, INDIA'S DIRECTOR GENERAL OF THE COUNCIL FOR SCIENTIFIC & INDUSTRIAL RESEARCH, TOLD *SCIENCE* MAGAZINE IN 2005.
Ireland	ALLOWS A 20 PERCENT R&D TAX CREDIT, A FULL DEDUCTION, AND A 12.5 PERCENT CORPORATE INCOME TAX RATE. CAPITAL EXPENDITURE MAY ALSO QUALIFY FOR A SEPARATE FLAT CREDIT.	NEARLY HALF OF THE COMPANIES SUPPORTED BY IRELAND'S INDUSTRIAL DEVELOPMENT AGENCY OUTLAY R&D EXPENDITURES.
Japan	A FLAT 10 PERCENT R&D TAX CREDIT (15 PERCENT FOR SMALL COMPANIES), AMONG OTHER INCENTIVES.	IN 2003, US SUBSIDIARIES SPENT US$1.7 BILLION ON R&D IN JAPAN. ACCORDING TO JAPAN'S MINISTRY OF ECONOMY, TRADE AND INDUSTRY, R&D AND IT TAX RELIEF HAS CREATED 400,000 JOBS AND BOOSTED GROSS DOMESTIC PRODUCT BY US$55 BILLION OVER A RECENT THREE-YEAR PERIOD.
South Korea	TAX HOLIDAYS UP TO SEVEN YEARS FOR HIGH-TECH BUSINESSES. TAX CREDITS FOR R&D EXPENDITURES.	SOUTH KOREA PROVIDES GENEROUS TAX INCENTIVES AND OTHER BENEFITS FOR FOREIGN COMPANIES LOCATING IN THE INCHEON FREE ECONOMIC ZONE.
Singapore	FOREIGN COMPANIES RECEIVE A FIVE-YEAR TAX HOLIDAY ON R&D INCOME.	SINGAPORE SEEKS AND NURTURES BUSINESS IDEAS OF ANY SIZE—EVERYTHING FROM STARTUPS WITH LITTLE MORE THAN THE GERM OF AN IDEA TO GLOBAL CORPORATIONS, SAYS A GOVERNMENT WEBSITE.
UK	ALLOWS A 125 PERCENT DEDUCTION (EQUAL TO A FLAT 7.5 PERCENT TAX CREDIT) FOR R&D EXPENSES, PLUS A 175 PERCENT DEDUCTION FOR EXPENDITURES EXCEEDING A BASE AMOUNT OF PRIOR-YEAR SPENDING.	THE UK LEADS THE WORLD IN ATTRACTING US R&D INVESTMENTS OVERSEAS. US SUBSIDIARIES SPENT MORE THAN $4 BILLION ON UK-BASED R&D IN 2003.

10.3 / RESEARCH AND DEVELOPMENT (R&D) TAX INCENTIVES BY COUNTRY

Tax incentives are used worldwide to encourage research, as summarized in the chart on the previous page from the 2007 "Rising Above the Gathering Storm" report commissioned by the National Academy of Sciences, the National Academy of Engineering, and the Institute of Medicine.

Warren Buffett's idea is to take "3 percent or something like a corporate income tax totally devoted to a fund that would be administered by some representatives of corporate America to be used in intelligent ways for the long-term benefit of society. This group could tackle education, health, et cetera, or other activities in which government has a large role. This would be, perhaps, $30 billion a year; you would exempt small companies."

Commit Globally to Environmental Research and Sustainability /

Climate change and pollution demand that the world commit wholeheartedly to sustainable research and development. The world's environmental woes cannot be solved in isolation. Even if the US completely eliminated its environmental impact on the planet, the country would still suffer the dire consequences of ill-managed growth in China, India, and elsewhere.

Economist Jeffrey Sachs, director of Columbia University's Earth Institute, states that **sustainable development is the key to easing global poverty** without further burdening the environment. Furthermore, addressing environmental problems will go a long way toward reducing global strife and warfare, he says. Sachs maintains that severe environmental conditions, such as worsening droughts, are at the root of violence and poverty in sub-Saharan Africa and regions in the Middle East and Central Asia.

"Look closely at the violence in Afghanistan, Chad, Ethiopia, Iraq, Pakistan, Somalia, and Sudan—one finds tribal and often pastoralist communities struggling to survive deepening ecological crises. Water scarcity, in particular, has been a source of territorial conflict when traditional systems of land management fail in the face of rising populations and temperatures and declining rainfall," Sachs wrote in the February 2008 edition of *Scientific American*.[4] "Washington looks at many of these clashes and erroneously sees Islamist ideology at the core ... the root of the crisis in the dryland countries is not Islam but extreme poverty and environmental stress."

Rather than military interventions, what's needed are environmental measures such as rainwater harvesting and underground aquifers. Population increases and climate change will intensify water stress throughout more regions. In his groundbreaking book, *The End of Poverty: The Economic Possibilities of Our Time*, Sachs calls for the following global environmental research and development initiatives.[5]

Tropical agriculture / New seed varieties and new water and soil management techniques.

Energy systems in remote areas / Technologies for off-grid power, including renewable energy sources (such as photovoltaic cells), power generators, improved batteries, and low-watt illumination.

Water management / Improved technologies for water harvesting, desalination and small-scale irrigation, and improved management of overused aquifers.

Sustainable management of ecosystems / Care of fragile ecosystems (coral reefs, mangrove swamps, fisheries, rainforests, etc.), which are succumbing to anthropogenic forces with dire consequences.

Climate forecasting and adjustment / Improved measurement of seasonal, inter-annual, and long-term climate changes, with a view toward adjustment to climate changes.

The Earth Institute maintains that the interconnected challenges of climate, environmental management, conservation, public health, and economic development must be addressed holistically. To that end, the institute has linked its key department to better carry out the following initiatives:[6]

Pioneering the use of geographic information systems in rural Ethiopia to track and respond rapidly to malaria outbreaks.

Using specially programmed cell phones in remote areas of Rwanda to provide real-time health data to the ministry of health.

Introducing agro-forestry techniques to triple food crops in nitrogen-depleted soils of Africa.

Designing efficient, low-cost batteries to power light bulbs in villages too remote to join a power grid.

Demonstrating how high-tech forecasting of El Niño fluctuations can help countries better time crop planting and harvesting and better manage water reservoirs and fisheries.

Using cutting-edge hydrology, geochemistry and public-health techniques to address arsenic poisoning in Bangladesh's water supply.

Ensuring global environmental responsibility is one of the eight Millennium Development Goals (MDG) that 189 UN member states pledged in 2000 to reach by 2015. Other goals target poverty, hunger, and disease. By 2008, at the halfway point of the deadline, The World Bank reported that progress for most of the goals was not on track, and that urgent action was needed on climate change.[7] The 2008 MDG Global Monitoring Report argued that environmental sustainability underpins progress on other MDGs. If forests are lost, soils degraded, and water and air polluted, and greenhouse gas emissions not contained, achievements in poverty reduction and human development will not be sustainable.

One of the eight goals, poverty reduction, was on track at the halfway point. While per capita GDP growth in developing countries has eased poverty, those countries need help to achieve growth with environmental sustainability if they are not to lose hard-won gains. Developing countries will suffer most from climate change. Transition to climate-resilient and low-carbon growth will require financing and technology transfer to developing countries, the report urged.

Develop and Connect Research Facilities Worldwide /

The most fertile incubator by which to grow the new sciences is a globally connected web of sophisticated research facilities. Funding for new and enhanced facilities must be provided by an international community of donors, government leaders, and collaborators.

National Laboratories /

Many developed countries operate national laboratories—extensive, one-of-a-kind government-supported research facilities. Rapid developments in research and technology warrant many more such laboratories, in both developed and developing countries. These laboratories must be networked to share knowledge and speed the development of the new sciences. Moreover, equipment and resources in each facility should be available to outside research teams, both onsite and electronically.

A terrific model is Argonne National Laboratory's Advanced Proton Source near Chicago, mentioned in chapter 8 under virtual reality. The synchrotron light source produces the Western hemisphere's highest-energy, highest-brilliance X-ray beams, allowing scientists to explore the structure and function of materials in the center of the Earth and in outer space, and all points in between.

The 40 beam lines, each with customized X-ray beams, provide an exceptional range of experimental conditions at a single facility. In 2006 alone, more than 2,000 scientists from around the world migrated there to conduct research in the fields of materials science; biological science; physics; chemistry; environmental, geophysical, and planetary science; and X-ray instrumentation. Such research has vast applications: it can impact the evolution of combustion engines and microcircuits, aid in developing new pharmaceuticals, pioneer nanotechnologies in which scale is measured in billionths of a meter, and much more.

The Advanced Proton Source is a true collaboration. Academic and industrial partners build the beamlines (more are underway), and funding comes from the US Department of Energy and state governments, as well as industry and private universities. Researchers interact with colleagues from industry and national laboratories, exchanging ideas both formally and informally through collaborations, seminars, and impromptu discussions.

Another Argonne-related project, TeraGrid, is the world's most comprehensive cyberinfrastructure for open scientific research. Operated by the National Science Foundation, TeraGrid combines advanced resources at 11 partner sites (including Argonne) to create an integrated, persistent computational resource, including more than 100 discipline-specific databases, for the nation's science and engineering community. It offers more than 750 teraflops of computing capability and more than 30 petabytes of online and archival data storage, with rapid access over high-performance networks.

As put forth by the authors of "Rising Above the Gathering Storm," the US government needs a National Coordination Office for Advanced Research Instrumentation and Facilities (ARIF). Critical to research, ARIFs house interacting instruments and networks of sensors, databases, and cyberinfrastructure. In the past 20 years, eight Nobel Prizes in physics were awarded to inventors of new instruments, such as the electron and tunneling microscope, laser and neutron spectroscopy, particle detectors, and the integrated circuit.

Large and expensive, these high-powered instruments require federal support. To that end, the "Gathering Storm" committee recommended in 2007 that $500 million per year over the next five years be allocated for the construction and maintenance of such research facilities. Universities and the government's national laboratories should compete annually for the funding, the committee holds.

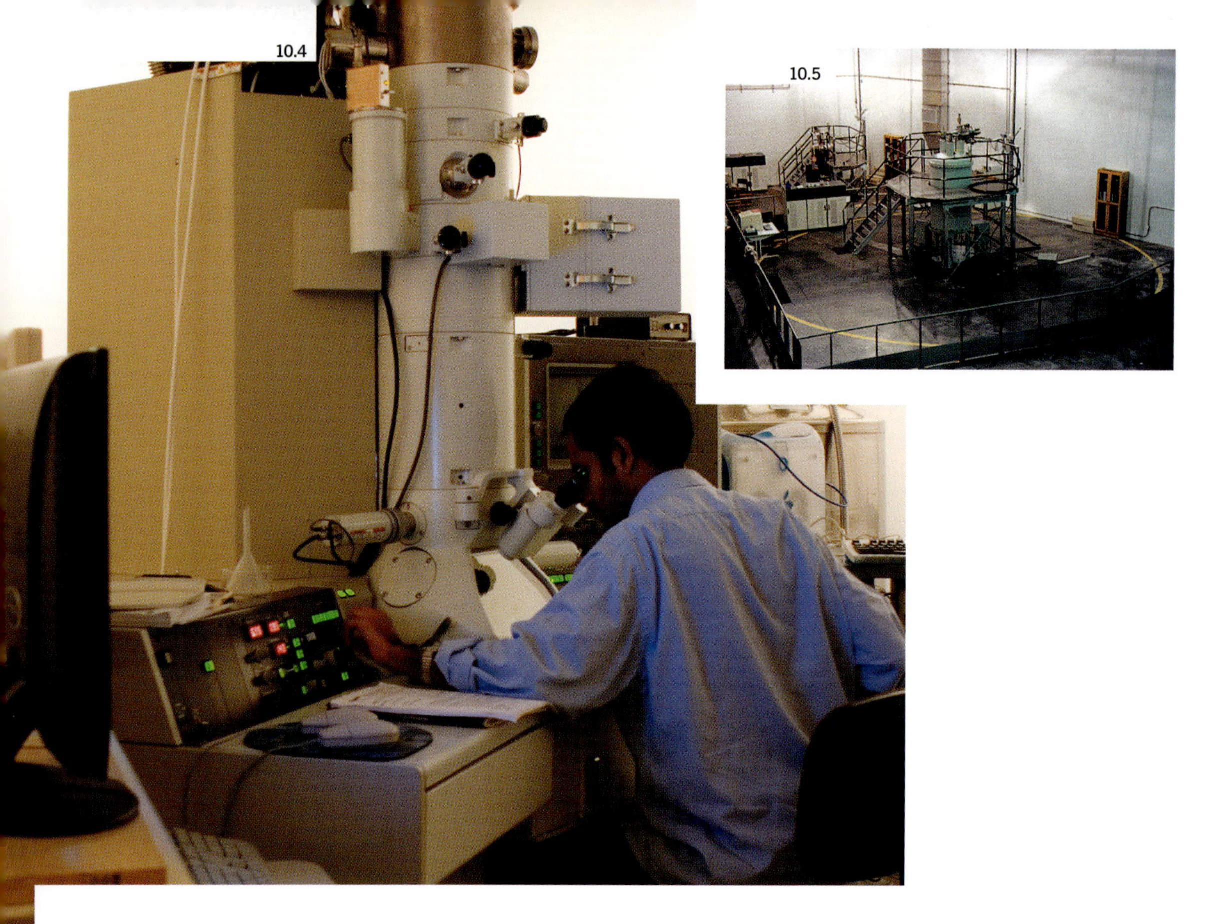

Research Parks and Towns /

As we increasingly appreciate both the interconnectivity of science and the power of collaboration, the environment in which we conduct research is changing accordingly. **Yesterday's research parks are becoming tomorrow's science towns**, uniquely structured to support interdisciplinary research and development on a world-class level within the confines, at least geographically, of a dedicated village.

Research parks, clusters of high-tech research, development and/or manufacturing concerns, are nothing new, of course. The world's first research park started in the early 1950s and foreshadowed what is known today as Silicon Valley. North Carolina's Research Triangle Park was another early innovation. In the ensuing decades, the world took notice of the wealth creation potential and technology-spurred growth of these clusters, especially Silicon Valley, and research parks became commonplace in the USA, Western Europe, and Japan.

Research parks foster innovation through knowledge partnerships among businesses, academia, and economic development concerns. They are best known for nurturing startup companies "spun out" from a university or company. But because the founding premise of collaboration and support has worked so well, the concept has broadened to include major, privately owned research organizations and even federal laboratories, along with top universities. The involvement of a research university is often key. Such a university attracts a diverse range of intellectual talent and resources to create, test, and openly disseminate new knowledge that encourages multi-disciplinary interactions.

Whether hosting large or small players, **research parks enable unique, symbiotic relationships that speed and enhance the development of all parties.** They also create new jobs, new industries, and new solutions. According to a 2007 study by Battelle Memorial Institute, a private non-profit applied-science and technology development company, nearly 800 US firms had graduated from park incubators in the previous five years, while only 13 percent failed. About one-quarter of those graduate companies remained in their park, while fewer than 10 percent left the region. **The study also found that each job created in a research park generates 2.57 local jobs.**[8]

In 2007 Battelle and the Association of University Research Parks (AURP) collected data from 134 parks in Canada and the USA. The data for a typical research park in the survey based on the median is listed below.[9]

Size	• 114 ACRES • SIX BUILDINGS • ONLY 30 PERCENT OF TOTAL ESTIMATED SQUARE FOOTAGE IS CURRENTLY DEVELOPED • 30,000 SQUARE FEET OF INCUBATOR SPACE
Location	• SUBURBAN COMMUNITY • POPULATION FEWER THAN 500,000 PEOPLE
Governance	OPERATED BY THE UNIVERSITY OR UNIVERSITY-AFFILIATED NON-PROFIT
Tenants	• 72 PERCENT ARE FOR-PROFIT COMPANIES • 14 PERCENT ARE UNIVERSITY FACILITIES
Employment	TYPICAL PARK EMPLOYS 750 PEOPLE
Major Industry Sectors	IT, DRUGS AND PHARMACEUTICALS, AND SCIENTIFIC AND ENGINEERING SERVICE PROVIDERS
Finances	• LESS THAN $1 MILLION-PER-YEAR OPERATING BUDGET • REVENUES PRIMARILY FROM PARK OPERATIONS BUT FINDS ALSO COME FROM UNIVERSITIES AND STATE, LOCAL, AND FEDERAL GOVERNMENT • LIMITED OR NO PROFITABILITY, 75 PERCENT OF THE PARKS HAVE NO RETAINED EARNINGS OF LESS THAN 10 PERCENT
Services	PROVIDE A RANGE OF BUSINESS AND COMMERCIALIZATION ASSISTANCE SERVICES, INCLUDING: • HELP IN ACCESSING STATE AND OTHER LOCAL PROGRAMS • LINKING TO OR PROVIDING SOURCES OF CAPITAL • BUSINESS PLANNING • MARKETING AND SALES-STRATEGY ADVICE • TECHNOLOGY AND MARKET ASSESSMENT

Today, many nations are seizing the concept of ultra-modern research parks as a prominent economic development instrument. "No matter how they are called, be it science parks, high-tech centers, incubator centers, technology parks, technoparks or science cities, they have given hopes to policymakers in many countries to boost regional technology transfer, innovativeness, and hence competitiveness," says a joint report of the UN Educational, Scientific and Cultural Organization and the World Technopolis Association (WTA).[10]

Recently, leading-edge planned research communities have taken the integration of science-and-technology research, business, academia, and government to an unprecedented level of whole-life integration of the people—the human capital—involved in interrelated endeavors.

10.4 / ELECTRON MICROSCOPES SUPPORT GRADUATE EDUCATION.

10.5 / TWO STATE-OF-THE-ART NMRS.

The justification of such a costly and complex development is that in today's knowledge economy, human capital reigns supreme over financial capital. In creating such a world-class research and development hub where international partnerships can flourish, the economic development and the international standing of the region is fueled.

"The key to regional growth lies not in reducing the costs of doing business, but in concentrating a more critical core of highly educated and productive people," concludes an extensive 2007 report by the Urban Research Program (URP) of Griffith University in Queensland, Australia. The US's technology-fueled economy in the 1990s is testament to that. "At the regional and/or local level the impact becomes more acute as a region's comparative advantage is now, more than ever, tied to its ability to attract the right talent," adds the URP report, titled "The Role of Community and Lifestyle in the Making of a Knowledge City."[11]

The key to creating a top-talent science community, therefore, lies in attracting the right people. To that end, no premium lifestyle amenity is spared. Beyond creating the best-in-class research infrastructure, today's science towns are creating idyllic, multi-cultural communities whereby work and play are fascinatingly intertwined. **The goal is nothing short of creating a knowledge-worker's paradise.**

Some of the characteristics of these communities include:

- Proximity of work, residential areas, and recreation. Walking and bicycling are easy options, although transit, local and regional, is convenient as well;
- A rich and internationally diverse assortment of retail, culinary, cultural, and entertainment offerings, including a vibrant nightlife, service the affluent customer base 24 hours a day, seven days a week;
- A commitment to sustainable living and integration with nature, including extensive green space, parks, bodies of water, and even old-growth trees;
- Residences, though varied, are typically smaller, and because knowledge workers are highly mobile, there is an ample supply of rental accommodation;
- Excellent educational institutions, both in the form of top-level university affiliation and primary and secondary schools;
- Connectivity in its highest, most convenient form;
- The opportunity to live and work among an internationally and scientifically diverse set of accomplished peers.

In a model research town, amenities such as housing, retail, schools, and recreation all center around research facilities. Traffic is minimal as researchers and other community members easily walk or bike to work and other destinations. Because everything is close at hand, researchers can take a midday break to exercise, relax, lunch with family, or tend to a simple chore. Researchers have more time to think, see what others are doing and share ideas outside the lab environment, all of which provides fertile ground for discoveries to be accelerated and life to be enjoyed to the fullest.

North Carolina: Research Triangle Park and Centennial Campus /

Research Triangle Park (RTP) was founded in 1959 by government, university, and business leaders as a model for research, innovation, and economic development. By establishing a place where educators, researchers and businesses could come together as collaborative partners, the founders hoped to change the economic composition of the region and state.

The vision was to provide a ready physical infrastructure that would attract research-oriented companies. The advantage of locating in RTP would be that companies could employ the highly educated local work force and utilize the rich research resources of nearby Duke University, NC State University, and the University of North Carolina at Chapel Hill.

RTP now comprises more than 7,000 acres, more than 25-million square feet of developed space and more than 170 organizations involved in research and development. Furthering the mission, NC State University, which anchors one corner of RTP, has been busily creating its vision of the future: Centennial Campus, a "technopolis" designed to advance technologies ranging from semiconductors to genomics. The 1,334-acre campus consists of multi-disciplinary research-and-development neighborhoods, with university, corporate, and government facilities intertwined. Housing and recreation are also included.

"We ought to have a place where we can have business and universities and the best thinkers all working together, working alongside each other, parking in the same parking lots, having lunch together, sharing ideas, and coming up with the best thoughts about how we can do things, how we can create new technologies, new companies, and new jobs."[12]

– FORMER NORTH CAROLINA GOVERNOR JAMES HUNT

NC State University's Centennial Campus was carved out in 1984 when former North Carolina Governor James Hunt transferred 355 acres of land to the university. The intent was to create a place where researchers, students, and private industry could partner to generate ideas that would strengthen the state's economy. The thriving research community now comprises more than 130 companies, government agencies, and university research and academic units. More than 1,600 corporate and government employees work alongside 1,350 faculty, staff, post-docs, and students. Major partners include ABB, Red Hat, the Iams Company, Ericsson IPI, the US Department of Agriculture, GlaxoSmithKline, MeadWestVaco, and Talecris Biotherapeutics. The National Weather Service also calls the campus home.

The Association of University Research Parks named Centennial Campus its 2007 Outstanding Research Park. "Centennial Campus exemplifies the new model of research science parks in which strategically planned mixed-use campus expansions create innovation, partnerships, and a high quality of life in their community and the nation," said AURP President Mike Bowman.[13]

10.6

About 2.7 million square feet of space on Centennial Campus is developed, and a total of 8 million square feet is planned over the next few decades. The campus will include a middle school, more residential housing, an executive conference center and hotel, a golf course, and a town center. Planners envisioned a true interactive campus where a researcher might leave her office, walk to the town center for dinner and return to her home overlooking Lake Raleigh, situated near the center of the campus.

In 1999, the university began relocating its College of Engineering from a variety of existing buildings on the main campus to Centennial Campus about 2 miles away. Two new engineering buildings are already in place. The motivation for moving the College of Engineering to Centennial Campus grew out of a desire to achieve:

- An increased focus on cross-disciplinary education for upper division programs;
- Consolidation and sharing of classrooms, courses, and laboratories;
- Support cross-disciplinary senior project designs;
- Facilitate undergraduate involvement in research;
- Increase educational partnership with industry;
- Integrate cross-disciplinary research activities.

The new College of Engineering will house the following departments: Chemical and Biomolecular Engineering, Materials Science and Engineering, Civil Engineering, Electrical and Computer Engineering, Computer Sciences Engineering, Industrial Engineering and Mechanical and Aerospace Engineering.

10.7

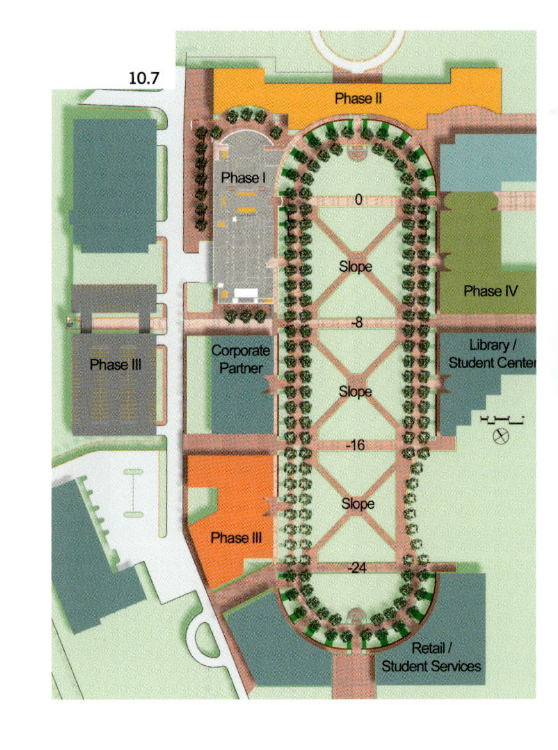

10.8

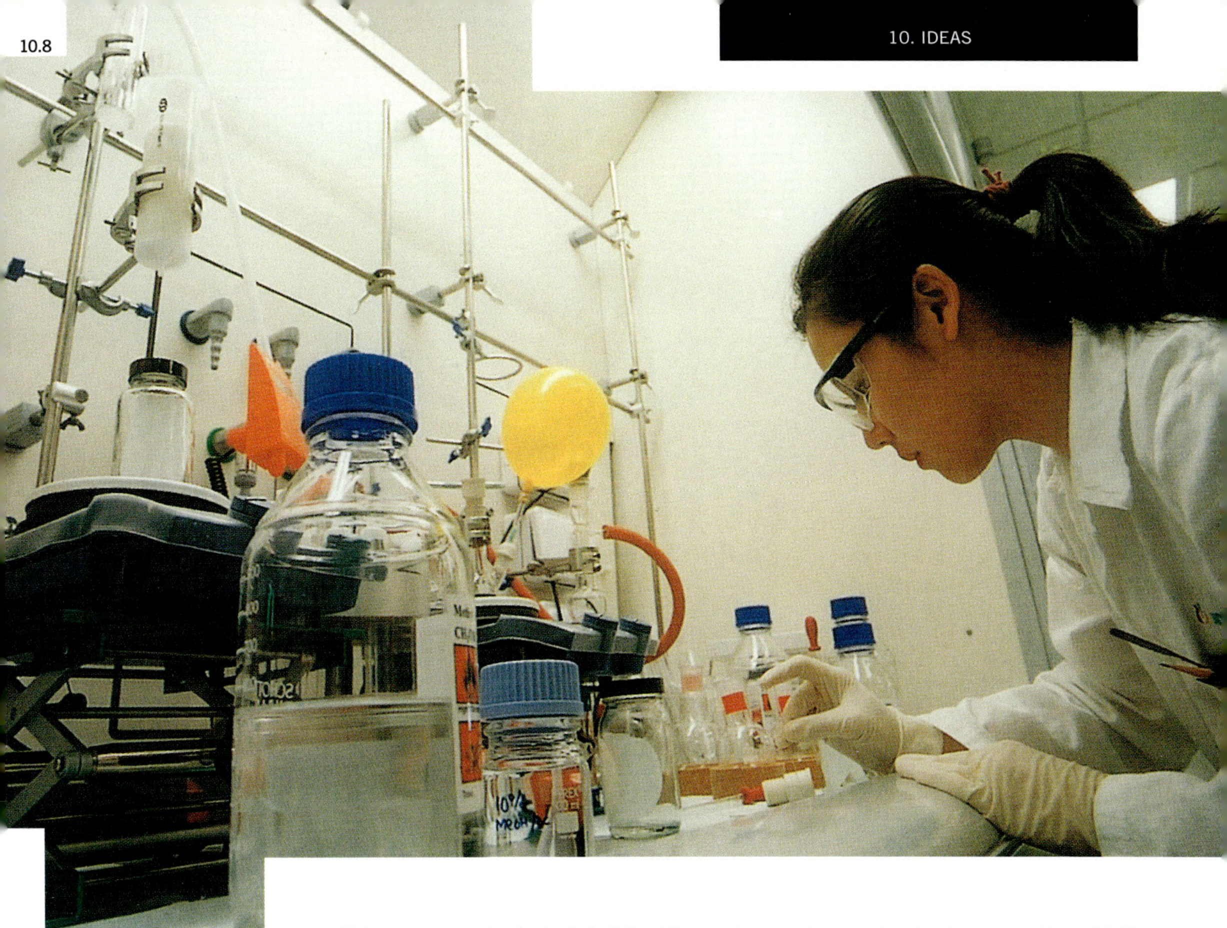

With an annual budget of $915 million and an endowment valued at more than $400 million, NC State University is a leader in non-federal-funded research, industry-funded research, and total expenditures for research and development. The school invests about $445 million a year in research, training, and extension, and faculty and students take part in more than 1,200 scientific, technological, and scholarly endeavors.

Technology transfer is a priority. The university is among the nation's top 20 for patents, with more than 500 issued, and for launching start-up companies based on university research. With students at the Centennial Campus, interacting with and learning from industry, government, and business partners every day, the school serves as a national model for how universities can construct highly effective academic partnerships with the private sector.

Singapore: Biopolis, Fusionopolis, and CREATE /

Southwest Singapore, home to the National University of Singapore, has become a world-class science hub. Major research centers, including Biopolis and Fusionopolis, enable a fusion of capabilities in line with multi-disciplinary research.

Begun in 2003, Biopolis is a biomedical sciences hub housing new and expanded research programs in genomics, molecular and cell biology, bioengineering and nanotechnology, medical biology, and clinical and translational research. Phase one, completed in 2004 at a cost of $500 million, consists of seven buildings linked by sky bridges to form a 185,000-square-meter complex housing up to 2,000 scientists. Phase two, a $70-million seven-story complex covering two blocks, is devoted to neuroscience and immunology. With its completion in late 2006, Biopolis grew to 222,000 square meters.

10.6 / TEACHING LABS REINFORCE INTERACTIONAL LEARNING.

10.7 / MASTER PLAN OF THE NEW COLLEGE OF ENGINEERING AT CENTENNIAL CAMPUS.

10.8 / RESEARCHER DOING STATE-OF-THE-ART WORK AT THE NEW BIOPOLIS CAMPUS.

10.9

10.9 / THE SEVEN BUILDINGS THAT MAKE UP PHASE ONE OF BIOPOLIS.

The nine buildings have been named according to their scientific endeavor.

Centros / A combination of the words "Centromere" and "Central." Centromere is the region of a chromosome vital to proper nuclear division and growth. The Centros building is occupied by the Agency for Science, Technology, and Research (A*STAR), the agency central to Singapore's development of biomedical sciences.

Chromos / Derived from the word "chromosome," a thread-like body usually consisting of DNA in a single long double helix containing genes. The name reinforces the idea of close interaction among tenants, much like the compact structure of chromosomes.

Genome / The total set of genetic information in an organism, it is an apt name for a building that houses the Genome Institute of Singapore, the center for genomic discovery and investigation of post-sequence genomics.

Helios / In Greek mythology, the sun god, who provides brilliant light and life-giving energy, is the namesake for this center of scientific research.

Matrix / A biology term for intercellular substances in which cells are embedded and grown. It is also a mathematical term referring to an arrangement of numbers and symbols on a grid.

Nanos / The Greek word for dwarf. In English "nano" denotes one-billionth of a unit. This building houses the cutting edge research conducted by the Institute of Bioengineering and Nanotechnology.

Proteos / From the word protein, present in all living things and central to many biological and cellular processes. An apt name for the Institute of Molecular and Cell Biology, which has core strengths in cell cycling, cell signaling, cell death, cell motility, and protein trafficking.

Immunos and Neuros / These two additional buildings, added as part of phase two, are devoted to immunology and neuroscience.

A government-led initiative, Biopolis is attracting overseas universities, research institutes, and business enterprises. **By sharing state-of-the-art facilities, scientific infrastructure, and specialized services, companies can significantly reduce research-and-development costs and accelerate timelines.**

Shared equipment includes machines for X-ray crystallography, nuclear magnetic resonance, electron microscopy, 9.4T MRI, and DNA sequencing. A research facility with specific

10.10 10.11 10.12 10.13

pathogen-free research animals is also available. **Principal investigators, enticed there by large five-year grants with no strings attached, say they can accomplish in Singapore in months what would take years elsewhere just to clear the red tape.**

Biopolis has made its mark in research areas such as tropical diseases and stem-cell development. Phase three, opened in 2009, is designed to attract medical-technology research companies to tap synergies within the Biopolis cluster.

Biopolis is a complete community replete with an oasis of retail, restaurant, and recreation facilities. The shady walkways, framed on all sides by gleaming hi-tech buildings housing biomedical companies, form the perfect setting for diners to linger over coffee after lunch or enjoy dinner under the stars.

Nearby Fusionopolis, conceived as a "world within a city," opened phase one early in 2008. Housing research organizations, high-tech companies, government agencies, retail outlets, and serviced apartments in one location, it provides a total environment in which to work, discover, create, play and bring ideas from mind to market. In addition to providing the latest scientific infrastructure and a test-bedding hub, the 120,000-square-meter two-tower complex also features a clubhouse, a rooftop swimming pool, a performance theater, and sky gardens.

The science-and-technology powerhouse will be built in five phases on 30 hectares. The $250 million Phase 2A complex houses dry and wet laboratories, clean-room facilities, and ground-floor retail units.

Fusionopolis will bring together 1,500 research scientists, engineers, and technology experts from the six of the seven public labs of A*STAR as well as the private-sector institutes conducting complementary research and development in programs such as infocomms and media, high-performance computing, manufacturing, microelectronics, data storage, and chemical and materials science. High-tech provisions will include an anechoic chamber, a research-and-development foundry, and a nanofabrication-and-characterization facility.

In further pursuit of its science-driven future, Singapore is also carving out a futuristic research town near Biopolis and Fusionopolis called **CREATE** (Campus for Research Excellence and Technological Enterprise), which will open in 2010. **The multi-national, multi-disciplinary research enterprise will stimulate innovation by facilitating interaction among scientists and engineers from around the globe.** Singapore's National Research Foundation plans for CREATE to house several top research centers, the first being a groundbreaking alliance with MIT.

10.10 / CHILD-CARE AND RECREATION FACILITIES ARE PROVIDED AT BIOPOLIS TO SUPPORT INDIVIDUALS AND THEIR FAMILIES.

10.11 / RESTAURANTS IN BIOPOLIS ARE CONVENIENTLY LOCATED AND HELP TO SUPPORT COLLABORATION.

10.12 / RETAIL AREAS IN BIOPOLIS MAKE SHOPPING CONVENIENT AND HELP TO CREATE A SMALL-TOWN ATMOSPHERE.

10.13 / PUBLIC SPACES AND TRANSPORTATION FOR SOCIAL ENGAGEMENT ARE A KEY AMENITY FOR THIS NEW INTERNATIONAL RESEARCH COMMUNITY.

10.14

The Singapore-MIT Alliance for Research and Technology (SMART) Center will be MIT's first research center of its kind outside of Cambridge, Massachusetts, and MIT's largest international research endeavor. The center, to house over 400 researchers, will serve as an intellectual hub for interactions between MIT and global researchers in Singapore. Following MIT's lead, the Swiss Federal Institute of Technology was the second world-class institution to announce plans to set up shop at CREATE.

Designed by Perkins+Will, CREATE is pioneering the use of advanced environmental and energy-saving technologies for the tropics. "This will be far more than a research complex in the tropics—it will be an integral part of the tropical ecosystem," reported Perkins+Will's Russ Drinker.

In the CREATE centers daylight will flood the interiors, while energy for electric lighting will be solar-generated. Green roofs will reduce stormwater runoff, minimize interior heat gain and conserve energy, while basic water needs will be met by capturing tropical rainfall. The landscape is modeled after a tropical rainforest, reinforcing the natural environment and helping recreate a habitat attractive to endangered animal species, leading to a potential rebirth of the site's natural ecosystem.

Saudi Arabia: Jeddah BioCity /

Jeddah BioCity, recently completed in Saudi Arabia, will make its mark as the Middle East's first highly advanced biotechnology complex. The 1-million-square-meter campus will be the region's largest scientific, medical, manufacturing, and economic center. Innovative research and development is expected to lead to marketable products and improve the welfare of area residents.

In partnership with the King Faisal Specialist Hospital and Research Center, the project's mission statement emphasizes transference of biotechnology, high economic return, and service to the community. BioCity will hasten the transference of technology and knowledge to Saudi companies and universities and help train a workforce to compete in the international biotechnology arena.

Switzerland: BioValley and BioAlps /

Founded in 1996, BioValley is the leading life-science cluster in Europe. It connects academia and companies of three Upper Rhine Valley nations—France, Germany, and Switzerland. The biocluster's main objective is to generate systematic cooperation among all those involved in the life-science sectors, including pharmacology, biotechnology, nanotechnology, medical technology, chemistry, and agricultural biotechnology. It also makes a point to involve numerous smaller enterprises and suppliers, as well as organizations involved in technology, finance, and economic development.

"The relatively small area houses operations of 40 percent of the world's pharmaceutical industry, including almost

10.15

400 biotechnology companies and more than 150 academic or public institutions," reported *Modern Drug Discovery* magazine.[14] "Fifteen thousand scientists populate the area along with 70,000 students, making BioValley one of the top three densest European bioregions, joining the Cambridge biocluster in eastern England and the Medicon Valley cluster in Copenhagen, Denmark, and Sweden."

BioValley has set up sophisticated networks to link, among others, researchers, students, and venture capitalists. Well-developed instruments facilitate technology transfer and encourage exchange and collaboration among all involved.

BioAlps' Lake Geneva BioCluster, a European biotech and biomed leader, involves about 500 biotech and medtech companies, 500 research laboratories, 18 research institutions, and several science parks and incubators for start-up companies. The Swiss government strongly backs biomedical research and development, the universities run excellent basic and applied research programs, and start-ups are well supported.[15]

Other large-scale research campuses around the world include BioDelta in The Netherlands, Tsinghua Science Park in China, and New Songdo City in South Korea.

The following are some examples to consider for improving existing research parks if you intend to strategically build a new research campus.

Salk Institute: San Diego, USA /

- Research icon (built in the 1960s)
- Architectural icon (Lou Kahn, architect)
- Interstitial concept
- Flexible design

10.14 / A BIRD'S-EYE VIEW OF CREATE IN SOUTHWEST SINGAPORE.

10.15 / SALK INSTITUTE.

10.16

10.17

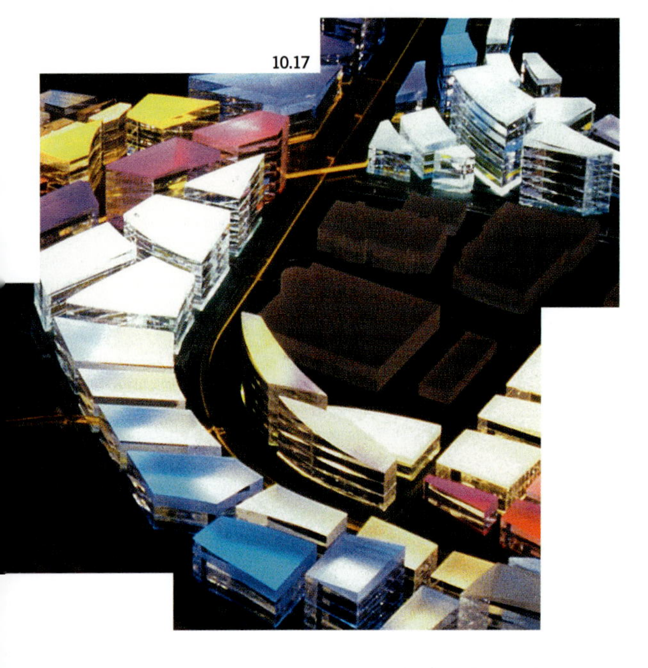

10.18

Sanger Center: Cambridge, UK /

- Site is in an area of natural beauty
- Multidisciplinary research
- Global research collaboration

Biopolis: Singapore /

- International brand to market researchers
- Government strategy for growth
- Shared facilities
- Campus amenities—retail, restaurants, recreation
- Sustainability

CREATE: Singapore /

- Research town
- Focus on improving research models, and collaboration
- Sustainability
- Basement as "research retail" for all

Medical University: Riyadh, Saudi Arabia /

- Includes on-site housing and transportation
- Modern architecture with local details
- Adaptive buildings
- Just-in-time services
- Sustainable

Quality to Compete /

Updating existing facilities or building new structures and research models should support blue-sky research. Truly state-of-the-art laboratories are being built throughout the world. Government agencies, private industry, and academia are building phenomenal buildings with ample natural indirect light, thoughtful interior design, inviting work spaces, support spaces to encourage interaction, and architectural and engineering systems to support change as the research itself changes. The best researchers are in high demand globally. In addition to offering outstanding research facilities, **competitive alliances are made to allow top researchers to partner with esteemed peers, either on the same campus or virtually.**

Blue-Sky Research /

We are in need of blue-sky research that inspires a global vision. The last time the USA shared such a vision was during John F. Kennedy's presidency nearly 50 years ago. The dream of sending a man to the moon was embraced by school children through to adults of all ages. We need similar visions today. In addition to the other research initiatives mentioned in this book, obvious candidates for blue-sky research include:

- Sustainable research to attain zero-carbon emissions
- Stem-cell research to increase life expectancy
- Nanotechnology to create super materials that are lighter, stronger, more efficient, and sustainable
- Genomics and proteomics to enable health issues to be mapped very early in life, allowing proactive health management.

Global collaboration is needed to accelerate these initiatives. Governments and corporations must explore ways to effectively collaborate to significantly **improve the "science of science."** Throughout this century research should drive industry and promote the healthy growth of developing countries, closing the huge life-expectancy gap between the poorest and richest nations. The research discoveries of the past 200 years should benefit people in all countries, and future discoveries should be shared with everyone. By the end of the 21st century, as discoveries continue and more information is shared, everyone around the world should have a reasonable expectation of living more than 80 years.

The proportion of research conducted by the USA compared to the rest of the world will decline slightly (even as US investments increase) because European countries will increase their investments. Most developing countries, especially China and India, will also invest more in research. We must maintain competition while strategically coordinating research projects.

In order for all of this to happen we require strong political leadership. Private industry and academia will be much more successful in supporting this initiative with thoughtful political strategies. The opportunities, and responsibility, to bring about positive changes belong to all of us, so we must accept the challenge and **insist on blue-sky leadership from politicians, and leaders of industry, and academia around the world.**

How much things improve will be determined by how people, companies, and governments work effectively together. Out of the 2008 global financial crisis has evolved the possibility, and the need, for many to re-evaluate how they do business in order to be more effective with less. The 21st century should be the century where we actually do the right things the right way for the global good.

10.16 / SANGER CENTER: CAMBRIDGE, UK.

10.17 / BIOPOLIS: SINGAPORE.

10.18 / CREATE: SINGAPORE.

10.19 / MEDICAL UNIVERSITY: RIYADH, SAUDI ARABIA.

RESOURCES + CREDITS

INTRODUCTION /

IMAGE CREDITS /

i.1 / ©John Glowczwski

i.2 / Photo courtesy of the Edison and Ford Winter Estates, ©Lisa Buttoni

i.3 / Photo courtesy of NIH, National Human Genome Research Institute (NHGRI)

i.4 / Photo courtesy of NIH, National Institute of General Medicine Sciences (NIGMS)

i.5 / Photo courtesy of the Wellcome Photo Library

i.6 / ©Yirui Sun

i.7 / Photo courtesy of the Wellcome Library, London

i.8 / ©2008–9 Kellar Autumn / KellarAutumn.com

CHAPTER 01. SCIENCE IS GOLDEN /

TEXT REFERENCES /

1. J.J. Jamrog, "The Perfect Storm: The Changing Nature of People," Human Resource Institute at the University of Tampa, September, 7, 2003.

2. T.L. Friedman, *The World is Flat, A Brief History of the Twenty-First Century*, Farrar, Straus and Giroux, New York, 2005, p. 8.

3. J.D. Sachs, *The End of Poverty, Economic Possibilities of Our Time*, Penguin Group (USA), New York, 2006, p. 16.

4. K. McCarthy, "Global Shifts in Population: The Coming Pressures of Immigration", RAND Corp, Santa Monica, 2001.

5. "People with Multiple Chronic Conditions Account for 65 Cents of Every Health Care Dollar", Partnership to Fight Chronic Disease, December 10, 2007.

6. P. Macinnis, *100 Discoveries: The Greatest Breakthroughs in History*, Metro Books, New York, 2008, pp. 234, 246, 257.

7. World Health Organization, *Preventing Chronic Diseases: A Vital Investment*, 2005.

8. Ibid.

9. D.A. Kessler, *The End of Overeating: Taking Control of the Insatiable American Appetite*, Rodale Books, New York, 2009.

IMAGE CREDITS /

01.1 / Photo courtesy of the Carter Center and the Ghana Guinea Worm Eradication Program, ©A. Poyo

01.2 / Photo courtesy of Dr. Walter Bradley, Baylor University; ©Elisa Guzman Teipel and Anna Morton, 2007

01.3 / Photo courtesy of Dr. Walter Bradley, Baylor University; ©Elisa Guzman Teipel, 2007

01.4 / Photo courtesy of Dr. Walter Bradley, Baylor University; ©Elisa Guzman Teipel, 2007

01.5 / Photo courtesy of Dr. Walter Bradley, Baylor University

01.6 / ©Michelle Litvin

01.7 / Photo courtesy of the Museum of Science and Industry

01.8 / Graph Created by Kalie Watch

01.9 / Graph created using statistics from: S. Moore, J.L. Simon, and the Cato Institute, "The Greatest Century That Ever Was: 25 Miraculous Trends of the Past 100 Years," policy analysis no. 364, December 15, 1999, pp. 1–32

01.10 / ©United Nations, 2009 World Mortality Chart

01.11 / Source: sciencemag.org

01.13 / Photo courtesy of the CDC, Chris Zahiser, B.S.N., R.N., M.P.H.

01.14 / Courtesy of the Wellcome Library, London

01.15 / Source: National Center for Health Statistics: National Vital Statistics Reports (53)5:Table 11

01.16 / Source: National Center for Health Statistics: National Vital Statistics Reports (53)5:Table 11

CHAPTER 02. INTERNATIONAL COLLABORATION /

TEXT REFERENCES /

1. UK Office of Science and Innovation, *International Research and Collaboration*, March 2007, p. 1.

2. National Science Foundation, *The Science and Engineering Workforce: Realizing America's Potential*, November 4, 2008, p. 1.

3. National Academy of Sciences, National Academy of Engineering, and Institute of Medicine, *Rising Above the Gathering Storm, Energizing and Employing America for a Brighter Economic Future*, National Academies Press, Washington, D.C., 2007, p. 3.

4. National Science Foundation, S&E Indicators 2006, "Science and Engineering Labor Force: Global S&E Labor Force and the United States," chapter 3.

5. Ibid., p. 1.

6. Ibid., conclusion.

7. "China's advanced technology state," *The Washington Times*, September 9, 2005, p. 4.

8. National Science Foundation, S&E Indicators 2006, "Higher Education in Science and Engineering: Global Higher Education in S&E," chapter 2.

9. D. Normile, "Overseas Insourcing," *Electronic Business*, July 2005.

10. B. Einhorn, J. Carey & N. Gross, "A New Lab Partner for the U.S.?" *Business Week*, August 22, 2005, p. 116.

11. "Pop-Up Cities: China Builds a Bright Green Metropolis," *Wired Magazine*, May 2007.

12. National Science Foundation, "SRS Asia's Rising Science and Technology Strength: Comparative Indicators for Asia, the European Union, and the United States," August 2007.

13. "2008-WCY-Rankings," <http://www.imd.ch/news/2008-WCY-Rankings.cfm>.

14. "Innovation News forecast," *Technology Review*, September 2004, p. 39.

15. National Science Foundation, S&E Indicators 2008, "Research and Development: National Trends and International Linkages, International R&D Comparisons," chapter 4, <http://www.nsf.gov/statistics/seind08/c4/c4s5.htm>.

16. "Investing in Research: 3 per cent of GDP," EurActiv.com, October 27, 2006, p. 2.

17. "Putting knowledge into practice: A broad-based innovation strategy for the EU," Commission of the European Communities, September 13, 2006, p. 2.

18. C. Bolgar, "Investing in France," France_wsje_0126, Special Advertising Section, January 23, 2006, p. 3.

19. National Academy of Sciences, National Academy of Engineering, and Institute of Medicine, *Rising Above the Gathering Storm, Energizing and Employing America for a Brighter Economic Future*, National Academies Press, Washington, D.C., 2007, p. 198.

20. J.D. Sachs, *The End of Poverty, Economic Possibilities of Our Time*, Penguin Group, New York, 2006, p. 350.

21. Ibid., p. 252.

22. Study South Africa, Research and Development, November 10, 2006, <http://www.dst.gov.za/centres-of-excellence>.

23. Ibid.

IMAGE CREDITS /

02.1 / Photo courtesy of Wellcome Trust, London

02.2 / Source: Science and Engineering Indicators 2008

02.3 / Source: Science and Engineering Indicators 2008

02.4 / ©Tim Griffith

02.5 / Photo courtesy of Biopolis, ©Ken Seet, Tay Kay Chin and Nicholas Leong

02.6 / Rendering by Perkins+Will

02.7 / Rendering by Perkins+Will

02.8 / Chart by Kalie Watch

CHAPTER 03. RESEARCH IN THE USA /

TEXT REFERENCES /

1. L. Iversen, *Drugs, A Very Short Introduction*, Oxford University Press, New York, 2001, pp. 7–10.

2. "Estimating the cost of new drug development: is it really 802 million dollars?," Health Aff (Millwood), March–April 2006, PubMed.gov.

3. L.D. Soloman, *The Quest for Human Longevity: Science, Business, and Public Policy*, Transaction Publishers, New Jersey, 2005, p. 14.

4. M.J. Lamberti (editor), *Clinical Trials State of the Industry Report 2007, The Pivotal Players, Pipelines and Perspective Driving Drug Development*, AHC Media LLC, Atlanta, 2007, p. 83.

5. Ibid., p. 5.

6. J. Schmidt, "'FDA caution, research drought cuts drug approvals'," *USA Today*, February 4, 2008, p. 1.

7. "Generic competition has negative fallout for Novartis bottom line," *The Shanghai Daily*, January 18, 2008.

8. L. Rapaport, "Wyeth Bad News in '07 Only Worsens on Drug Delays," *Bloomberg News*, New York, January 17, 2008.

9. G. Blumenstyk, "Northwestern U. Sells Royalty Rights From Blockbuster Drug for $700 Million," *The Chronicle of Higher Education*, December 19, 2007.

10. "Manhattan Drug Research Benefits University," *New York Times*, May 8, 2007.

11. J.E. Stiglitz, *Making Globalization Work*, W.W. Norton & Company, New York, 2006, pp. 122–3.

12. F. Zakaria, *The Post-American World*, Norton paperback, New York, 2009, p. 186.

IMAGE CREDITS /

03.1 / Hedrich Blessing, ©Nick Merrick

03.2 / Image courtesy of Oklahoma Medical Resaerch Foundation, Created by Timothy Mather, PhD.

03.3 / Image courtesy of Oklahoma Medical Resaerch Foundation, Created by Timothy Mather, PhD.

03.4 / Image courtesy of Oklahoma Medical Resaerch Foundation, Created by Timothy Mather, PhD.

03.5 / Image courtesy of Oklahoma Medical Resaerch Foundation, Created by Timothy Mather, PhD.

03.6 / Chart by Kalie Watch

03.7 / Chart by Kalie Watch

03.8 / Chart by Kalie Watch

CHAPTER 04. SUSTAINABLE SOLUTIONS /

TEXT REFERENCES /

1. T. Appenzeller, "The Big Thaw," *National Geographic*, June 2007, pp. 56–71.

2. B. Walsh, "Top Ten Green Ideas," *Time Magazine*, December 2007.

3. T. Macalister, "Investment fund giants demand 90% reduction in carbon emissions," *The Guardian*, February 29, 2008.

4. <http://www.usaid.gov/our_work/environment/water/case_studies/nyc.watershed.pdf>.

5. <http://www.treepeople.org/rainwater-resource>.

6. <http://www.planetizen.com/node/29166>.

7. M. Baum, *Green Building Research Funding: An Assessment of Current Activity in the United States*, U.S. Green Building Council, 2007, <http://www.usgbc.org/showfile.aspx?documentID=2465>.

8. Ibid., p. 1.

9. R. Gibson, *Moving Toward a Green Chemical Future*, California Research & Policy Center, July 2008.

10. M. LaMonica, "Green chemistry's 'race to innovation,'" *CNET News*, November 12, 2007, <http://www.tinyurl.com.au/2p2>.

11. "The Bioproducts Update: A green chemical revolution ...," Checkmate Public Affairs, October 10, 2006, <http://www.tinyurl.com.au/2p4>.

12. A. Waves, "The Future of Biofuels: a Global Perspective," *Economic Research Service*, November 2007.

IMAGE CREDITS /

04.1 / Photo courtesy of Getty Images (US), Inc., ©STR

04.2 / Created by Kalie Watch

04.3 / Courtesy of Perkins+Will, Peter Busby

04.4 / Courtesy of Perkins+Will, John Mlade

04.5 / Rendering by Perkins+Will

04.6 / Source: <http://energymap.mit.edu/>

04.7 / ©Michelle Litvin

04.8 / Courtesy of Perkins+Will

04.9 / Licensed from iStockphoto

04.10 / Licensed from iStockphoto

04.11 / Licensed from iStockphoto

04.12 / Photo courtesy of John Mlade

04.13 / Photo courtesy of NREL

04.14 / Photo courtesy of Suniva, Inc., ©Vinodh Chandrasekaran

04.15 / Photo courtesy of Biopolis; ©Ken Seet, Tay Kay Chin and Nicholas Leong

04.16 / Rendering by Perkins+Will

04.17 / Courtesy of Oklahoma Medical Research Foundation

04.18 / Photo courtesy of NREL

04.19 / Photo courtesy of NREL, ©Brent Nelson

04.20 / Photo courtesy of NREL, ©Warren Gretz

04.21 / Photo courtesy of NREL, ©Warren Gretz

04.22 / Photo courtesy of NREL, Sandia National Laboratories

04.23 / Photo courtesy of NREL, ©Robb Williamson

04.24 / Photo courtesy of NREL, ©Warren Gretz

CHAPTER 05. GLOBAL SUCCESS OF RESEARCH /

TEXT REFERENCES /

1. "An Overview of the Human Genome Project," National Human Genome Research Institute (NHGRI), <http://www.genome.gov/12011241>.

2. *Biology meets industry–genomics, proteomics, phenomics, Australian Academy of Science*, <http://www.science.org.au/nova/078/078print.htm>.

3. Ibid.

IMAGE CREDITS /

05.1 / Credit: Mehau Kulyk / Photo Researchers, Inc

05.2 / Image courtesy of Wellcome Images, ©Dr. TJ McMaster

05.3 / Image courtesy of Wellcome Images, ©Dan Salaman

05.4 / Image courtesy of Wellcome Images, The Sanger Institute

05.5 / Photo courtesy of the Automation Partnership

05.6 / Photo courtesy of the Automation Partnership

05.7 / Image courtesy of the Wellcome Library, London

05.8 / Image courtesy of Wellcome Images, ©Nicoletta Balayianni

05.9 / Image courtesy of Wellcome Images, ©Anton Enright

05.10 / Image courtesy of Wellcome Images, ©Neil Leslie

CHAPTER 06. POLITICAL SCIENCES /

TEXT REFERENCES /

1. R. Teixeira, "Public Opinion Snapshot: Solid Backing for Embryonic Stem Cell Research," Center for American Progress, March 30, 2007, <http://www.tinyurl.com.au/2p5>.

2. S.G. Stolberg, "Limits on Stem-Cell Research Re-emerge as a Political Issue," *New York Times*, May 6, 2004.

3. "The Michael J. Fox Foundation for Parkinson's Research–About The Foundation–About Michael–In His Own Words," <http://www.michaeljfox.org>.

4. "Stem Cell 101," Stem Cell Institute, University of Minnesota, <http://www.stemcell.umn.edu/>.

5. G. Vogel, "Researchers Turn Skin Cells Into Stem Cells," *ScienceNOW*, November 20, 2007, <http://news.sciencemag.org/sciencenow/2007/11/20-01.html>.

6. Ibid.

7. D. Bjerklie, "The Year In Medicine From A to Z," *Time*, November 26, 2006, <http://www.time.com/time/magazine/article/0,9171,1562958,00.html>.

8. "Future of Stem Cell Research–Video," globalchange.com, May 2004, <http://www.globalchange.com/stemcells2.htm>.

9. "Key Moments in the Stem-Cell Debate," NPR, November 20, 2007, <http://www.npr.org/templates/story/story.php?storyId=5252449>

10. "Stem Cells in Regenerative Medicine 2008 Market Report," Research And Markets, <http://www.researchandmarkets.com>.

11. "Asia Is Stem Cell Central," *BusinessWeek*, January 10, 2005, <http://www.businessweek.com/magazine/content/05_02/b3915052.htm>.

12. W. Arnold, "Singapore Acts as Haven for Stem Cell Research," *New York Times*, August 17, 2006, <http://www.tinyurl.com.au/2p8>.

13. H. Mehta, "India plans national stem cell initiative," SciDev.Net, November 24, 2005.

14. D. Normile, "Stem Cell Research: South Korea Picks Up the Pieces," *Science*, June 2, 2006, <http://www.sciencemag.org/cgi/content/full/312/5778/1298>.

15. "Worldwide HIV & AIDS Statistics," AVERT, <http://www.avert.org/worldstats.htm>.

16. "Priority environment and health risks," WHO, <http://www.who.int/heli/risks/en/>.

17. "Countries move toward more sustainable ways to roll back malaria," WHO, May 6, 2009, <http://www.tinyurl.com.au/2pd>.

18. J.D. Sachs, "The Neglected Tropical Diseases," *Scientific American*, January 2007, <http://www.sciam.com/article.cfm?id=the-neglected-tropical-diseases>.

19. "Cumulative Number of Confirmed Human Cases of Avian Influenza A/(H5N1) Reported to WHO," WHO, <http://www.tinyurl.com.au/2ph>.

20. "A Killer Flu?," Trust for America's Health, June 2005, <http://healthyamericans.org/reports/flu/>.

21. "International Sources of Assistance," WHO, <http://whqlibdoc.who.int/publications/2004/9241546158_chap6.pdf>.

22. "Deciphering Pathogens, Blueprints for New Medical Tools," U.S. Department of Health and Human Services, NIH Publication, September 2002.

23. C. Franco, "Billions for Biodefense: Federal Agency Biodefense Funding, FY2007–FY2008," Center for Biosecurity, UPMC, *Biosecurity and Bioterrorism*, vol. 7, no. 3, 2009, <http://www.tinyurl.com.au/2pi>.

24. "HHS Fact Sheet: Biodefense Preparedness: Record of Accomplishment," U.S. Department of Health and Human Services, April 28, 2004, <http://www.hhs.gov/news/press/2004pres/20040428.html>.

IMAGE CREDITS /

06.1 / Image courtesy of Wellcome Images, ©Karin Hing

06.2 / Image courtesy of Wellcome Images, ©Yirui Sun

06.3 / Image courtesy of Wellcome Images, ©Steven Pollard

06.4 / Image courtesy of Wellcome Images, ©Yirui Sun

06.5 / Graphic created by Patty Gregory

06.6 / Photo courtesy of the Wellcome Library, London

06.7 / Image courtesy of NIH, National Institute of Allergy and Infectious Diseases (NIAID)

06.8 / Image courtesy of CDC, ©C.S. Goldsmith and A. Balish

06.9 / Image courtesy of CDC, ©James Gathany

06.10 / Hedrich Blessing, ©Nick Merrick

06.11 / Hedrich Blessing, ©Nick Merrick

06.12 / Hedrich Blessing, ©Nick Merrick

06.13 / ©Michelle Litvin

06.14 / Photo courtesy of Perkins+Will

06.15 / Photo courtesy of CDC, ©James Gatheny

06.16 / Hedrich Blessing, ©Nick Merrick

06.17 / Image courtesy of CDC, ©Frederick Murphy

CHAPTER 07. NEW TECHNOLOGY /

TEXT REFERENCES /

1. "Nanotechnology," EU–European Information on Science & Research, EurActiv.com, <http://www.tinyurl.com.au/2pm>.

2. T. Theis, "Nanotechnology: A Revolution in the Making," IBM Research, symposium at Stanford University, July 19, 2001, <http://www.tinyurl.com.au/2pn>.

3. "Nanotechnology and Cancer: Fact Sheet", National Cancer Institute, <http://ncl.cancer.gov/ncl_business_plan.pdf>.

4. R. Langer, "The Nanotechnology Revolution," *The Futurist*, March-April 2006, p. 42.

5. "Nanoparticle Assembly Enters the Fast Lane," nanotechwire.com, October 11, 2006, <http://www.tinyurl.com.au/2pr>.

6. "Current Uses, Nanostructured Materials," Nanotechnology Now, March 13, 2006, p. 4, <http://www.nanotech-now.com/current-uses.htm>.

7. "Russia Commits 'Significant Resources' to Nano Research Through 2015", NanoScienceWorks, January 17, 2008, <http://www.tinyurl.com.au/2pt>.

8. *National Nanotechnology Initiative Strategic Plan*, Nanoscale Science, Engineering, and Technology Subcommittee, National Science and Technology Council, December 2004, p. iv.

9. K. Schmidt, *NanoFrontiers: Visions for the Future of Nanotechnology*, The Project on Emerging Nanotechnologies, March 2007, <http://www.tinyurl.com.au/2ui>.

10. Computational Neuroscience, Neuroscience Research Center, University of Texas Health Science Center at Houston, May 31, 2006, <http://www.cnsorg.org/index.shtml>.

11. K. Chaundy, "Why is Neuroscience Important?," Neuroscience Research Institute Ottawa, Canada, August 25, 2006, <http://www.nri.on.ca/home/index.html>.

IMAGE CREDITS /

07.1 / Image courtesy of Georgia Institute of Technology, Dr. Z.L. Wang

07.2 / Image courtesy of Georgia Institute of Technology, Dr. Z.L. Wang

07.3 / Image courtesy of Wellcome Images, ©Oliver Burston

07.4 / Image courtesy of Georgia Institute of Technology, Dr. Z.L. Wang

07.5 / Graphic created by Kalie Watch

07.6 / Image courtesy of Wellcome Images, ©Annie Cavanaugh

07.7 / Image courtesy of Wellcome Images, ©Annie Cavanaugh

07.8 / Image courtesy of Wellcome Images, ©Annie Cavanaugh

07.9 / Image courtesy of Wellcome Images, ©Oliver Burston

07.10 / Image courtesy of Georgia Institute of Technology, Dr. Z.L. Wang

07.11 / Image courtesy of Georgia Institute of Technology, Dr. Z.L. Wang

07.12 / Image courtesy of Georgia Institute of Technology, Dr. Z.L. Wang

07.13 / Image courtesy of Georgia Institute of Technology, Dr. Z.L. Wang

07.14 / Image courtesty of NIH, National Institute on Drug Abuse

07.15 / Image courtesy of NIH, National Institute of Mental Health

07.16 / Image courtesy of Wellcome Images, ©Benedict Campbell

07.17 / Image courtesy of Wellcome Images, ©Mark Lythgoe and Chloe Hutton

CHAPTER 08. IMPROVING THE SCIENCE OF SCIENCE /

TEXT REFERENCES /

1. "Stanford launches interdisciplinary initiative in the biological sciences," *Stanford Online Report*, June 9, 1999, <http://news.stanford.edu/news/1999/june9/biox-69.html>.

2. "Seasoned chef brings world cuisines to Clark Center restaurant," *Stanford Report*, Oct 22, 2003, <http://news.stanford.edu/news/2003/october22/xlinx-1022.html>.

3. "Unifying Science with Contemporary Design," R&D Magazine, May 18, 2004, <http://www.tinyurl.com.au/2q1>.

4. "Enhancing Science and Engineering at Harvard," Harvard, <http://www.provost.harvard.edu/reports/final_report.pdf>.

5. "Harvard alters its approach to scientific study," *The Boston Globe*, January 19, 2007, <http://www.tinyurl.com.au/2q4>.

6. Ibid.

7. Ibid.

8. Ibid.

9. "National Institute of Environmental Health Sciences Health Disparities Strategic Plan," <http://www.tinyurl.com.au/2q6>.

10. Ibid.

11. Ibid.

12. "Good News On Heart Attack And Chest Pain," *ScienceDaily*, May 2, 2007, <http://www.sciencedaily.com/releases/2007/05/070501160712.htm>.

13. "Deaths from heart attacks drop," *USA Today*, May 2, 2007, <http://www.tinyurl.com.au/2q7>.

14. Ibid.

15. American Psychiatric Association, <http://www.psych.org/>.

16. "Robotic Surgery", *ScienceDaily*, August 6, 2006, <http://www.sciencedailey.com>.

IMAGE CREDITS /

08.1 / ©Robert Canfield

08.2 / Chart created by Kalie Watch

08.3 / ©Robert Canfield

08.4 / ©Robert Canfield

08.5 / Created by Perkins+Will

08.6 / ©Robert Canfield

08.7 / ©Robert Canfield

08.8 / Image courtesy of the Wellcome Library, London

08.9 / Hedrich Blessing, ©Nick Merrick

08.10 / ©Michelle Litvin

08.11 / ©Michelle Litvin

08.12 / ©Michelle Litvin

08.13 / ©Michelle Litvin

08.14 / Graphic created by Kalie Watch

08.15 / Image courtesy of Patrick Bordnick

08.16 / Image courtesy of Patrick Bordnick

08.17 / Graphic created by Patty Gregory

08.18 / Image courtesy of Patrick Bordnick

08.19 / Image courtesy of Patrick Bordnick

08.20 / Image courtesy of Patrick Bordnick

08.21 / Image courtesy of Patrick Bordnick

08.22 / Image courtesy of University of Tennessee - Chattanooga

08.23 / ©Dan Watch

08.24 / ©Dan Watch

08.25 / ©Dan Watch

08.26 / ©Dan Watch

08.27 / ©Dan Watch

08.28 / ©Dan Watch

CHAPTER 09. EDUCATION + PHILANTHROPY /

TEXT REFERENCES /

1. E. Mansfield, "Academic research and industrial innovation," *Research Policy*, vol. 20, 1991.

2. "PISA 2006 results," OECD, <http://www.tinyurl.com.au/2qc>.

3. "Education at a Glance 2007: OECD Briefing Note for the United States," OECD, 2007, <http://www.oecd.org/dataoecd/22/51/39317423.pdf>.

4. "Launch of PISA 2006," OECD, <http://www.tinyurl.com.au/2ru>.

5. "Asia's Rising Science and Technology Strength: Comparative Indicators for Asia, the European Union, and the United States," National Science Foundation, *SRS Publications*, August 2007, <http://www.nsf.gov/statistics/nsf07319/>.

6. "Gathering Storm Gains Momentum," The National Academies, *In Focus*, vol. 6, no. 2, 2006, <http://infocusmagazine.org/6.2/20_20.html>.

7. E. Ramirez, "Room to Improve," *U.S. News & World Report*, November 2, 2007, <http://www.tinyurl.com.au/2rx>.

8. Ibid.

9. Ibid.

10. Ibid.

11. E. Chute, "How to take a course at MIT free–at home," *Pittsburgh Post-Gazette*, November 18, 2007, <http://www.tinyurl.com.au/2ry>.

12. The My Hero Project: MIT Open Courseware Project, MIT, <http://www.tinyurl.com.au/2s0>.

13. "iCampus: The Education Revolution," iCampus News, <http://icampus.mit.edu/news/index.shtml>.

14. Ibid.

15. Ibid.

16. "iLab: Remote Online Laboratories," iCampus Projects, <http://icampus.mit.edu/projects/iLabs.shtml>.

17. "Mi Lab Es Su Lab," Inside Higher Ed: News, March 24, 2006, <http://insidehighered.com/news/2006/03/24/ilab>.

18. "MISTI helps bring iLabs to China," MIT News Office, June 12, 2006, <http://www.tinyurl.com.au/2s3>.

19. "MIT iCampus Project Yields Key Educational Learnings," iCampus News, <http://www.tinyurl.com.au/2s5>.

20. M. Trumbull, "A new era for supercharged philanthropy," *The Christian Science Monitor*, June 28, 2006, <http://www.tinyurl.com.au/2s6>.

21. M. Jones, "In Lean Times, Scientists May Eye Funding From Private Foundations and Prizes," The Samuel Roberts Noble Foundation, May 12, 2008, <http://www.tinyurl.com.au/2s9>.

22. "How they give back," *Fortune*, CNNMoney.com, <http://www.tinyurl.com.au/2sa>.

23. "Philanthropy and the funding of science," *Current Science*, vol. 83, no. 5, September 10, 2002.

24. J.T. Bruer & S.M. Fitzpatrick, "Science Funding and Private Philanthropy," *Science*, vol. 277, no. 5326, August 1997, <http://www.tinyurl.com.au/2sc>.

25. B. Bergstein, "Philanthropy strives to leave mark on science," Technology & Science, msnbc, November 13, 2006, <http://www.msnbc.msn.com/id/15701654/>.

26. "Events in Philanthropy," naturenews, <http://www.tinyurl.com.au/2se>.

27. B. Tansey, "Silicon Valley venture capitalist backs offbeat cancer cure ideas," *San Francisco Chronicle*, January 15, 2008, <http://www.tinyurl.com.au/2sf>.

28. J.T. Bruer & S.M. Fitzpatrick, "Science Funding and Private Philanthropy," *Science*, vol. 277, no. 5326, August 1997, <http://www.tinyurl.com.au/2sc>.

29. Ibid.

IMAGE CREDITS /

09.1 / ©Michelle Litvin

09.2 / Image courtesy of Georgia Institute of Technology Center for Biologically Inspired Design, Jeannette Yen

09.3 / Rendering by Perkins+Will

09.4 / Rendering by Perkins+Will

09.5 / ©Terrie Watch

CHAPTER 10. IDEAS /

TEXT REFERENCES /

1. D. Tapscott, & A.D. Williams, *Wikinomics*, Penguin Group, New York, 2006, p. 17.

2. "Mission & History," TB Alliance, <http://www.tballiance.org/about/mission.php>.

3. Bio-IT World.com, <http://www.bio-itworld.com/>.

4. J.D. Sachs, "Crisis in the Drylands," *Scientific American*, February 2008.

5. J.D. Sachs, *The End of Poverty: Economic Possibilities for Our Time*, Penguin Books, New York, 2006, p. 283.

6. Ibid., p. 224.

7. "Global Monitoring Report Warns on MDG Goals," News & Broadcast, The World Bank, April 8, 2008, <http://www.tinyurl.com.au/2sj>.

8. "University research parks contribute to economic competitiveness of regions, states and nations," Association of University Research Parks, October 26, 2007, <http://www.tinyurl.com.au/2sk>.

9. Ibid.

10. "University-Industry Partnerships (UNISPAR)," Science Policy and Sustainable Development, UNESCO, http://www.tinyurl.com.au/2sn.

11. "Innovative Culture with Excellent R&D and Infrastrucure," For Investors, Australian Trade Commission, <http://www.tinyurl.com.au/2so>.

12. K. Nichols, "Centennial Campus Honored as Top Research Science Park," CSC News, NC State *Computer Science*, November 7, 2007, <http://www.csc.ncsu.edu/news/608>.

13. Ibid.

14. A. Byrum, "Bio-Valley", *Modern Drug Discovery*, December 2004.

15. BioAlps, <http://www.bioalps.org/>.

IMAGE CREDITS /

10.1 / Image courtesy of the Carter Center and Ethiopia Malaria Control Program, ©The Carter Center/Louise Grubb

10.2 / Image courtesy of the Carter Center and Ethiopia Malaria Control Program, ©The Carter Center/Louise Grubb

10.3 / Source: "Survey of Current Business," Bureau of Economic Analysis, Department of Commerce

10.4 / ©Michelle Litvin

10.5 / ©Dan Watch

10.6 / ©Michelle Litvin

10.7 / Rendering by Perkins+Will

10.8 / Photo courtesy of Biopolis, ©Ken Seet, Tay Kay Chin and Nicholas Leong

10.9 / Photo courtesy of Biopolis, ©Ken Seet, Tay Kay Chin and Nicholas Leong

10.10 / Photo courtesy of Biopolis, ©Dan Watch

10.11 / Photo courtesy of Biopolis, ©Dan Watch

10.12 / Photo courtesy of Biopolis, ©Ken Seet, Tay Kay Chin and Nicholas Leong

10.13 / Photo courtesy of Biopolis, ©Ken Seet, Tay Kay Chin and Nicholas Leong

10.14 / Rendering by Perkins+Will

10.15 / ©Gary McNay

10.16 / ©John Freebrey

10.18 / Photo courtesy of Biopolis, ©Ken Seet, Tay Kay Chin and Nicholas Leong

10.18 / Rendering by Perkins+Will

10.19 / Rendering by Perkins+Will

BIBLIOGRAPHY

Abramson, M.D., John, *Overdosed America*, Harper-Collins Publishers, New York, USA, 2004.

Fishman, Ted C., *China, Inc., How the Rise of the Next Superpower Challenges America and the World, Scribner*, New York, USA, 2005.

Friedman, Thomas L., T*he World is Flat, A Brief History of the Twenty-First Century*, Farrar, Straus and Giroux, New York, USA, 2005.

From the Editors of Scientific American, *Understanding Nanotechnology*, Time Warner Book Group, New York, USA, 2002.

Gingrich, Newt, *Winning the Future, A 21st Century Contract with America*, Regnery Publishing, Inc., Washington, D.C., USA., 2006.

Goozner, Merrill, *The $800 Million Pill, The Truth Behind the Cost of the New Drugs*, University of California Press, Berkeley and Los Angeles, USA, 2004.

Gorman, Kevin J. (editor), Putnam Associates, *The First Twenty Years Are the Hardest: Two Decades in Healthcare Strategy*, Bridgeway Books, Austin, USA, 2009.

Jones, Van, *The Green Collar Economy: How One Solution Can Fix Our Two Biggest Problems*, HarperCollins Publishers, New York, USA, 2008.

Kinsley, Michael (editor) with Connor Clarke, *Creative Capitalism, A Conversation with Bill Gates, Warren Buffett and other Economic Leaders*, Simon & Schuster, New York, USA, 2008.

Lamberti, Mary Jo (editor), *Clinical Trials State of the Industry Report 2007, The Pivotal Players, Pipelines and Perspective Driving Drug Development*, AHC Media LLC, Atlanta, USA, 2007.

Macinnis, Peter, *100 Discoveries, The Greatest Breakthroughs in History*, Metro Books, New York, USA, 2008.

National Academy of Sciences, National Academy of Engineering, and Institute of Medicine, *Rising Above the Gathering Storm, Energizing and Employing America for a Brighter Economic Future*, National Academies Press, Washington, D.C., USA, 2007.

Pickens, T. Boone, *The First Billion is the Hardest*, Crown Business, New York, USA, 2008.

Pisano, Gary P., *Science Business, The Promise, The Reality, and The Future of Biotech*, Harvard Business School Press, Boston, USA, 2006.

Sachs, Jeffrey D., *The End of Poverty: Economic Possibilities for Our Time*, Penguin Books, USA, 2006.

Soloman, Lewis D., *The Quest for Human Longevity, Science, Business, and Public Policy*, Transaction Publishers, New Brunswick, USA, 2006.

Steer, Dr. Mark, Haley Birch, & Dr. Andrew Impney (editors), *Defining Moments in Science, Over a Century of the Greatest Discoveries, Experiments, Inventions, People, Publications, and Events that Rocked the World*, Cassell Illustrated, London, England, 2008.

Stiglitz, Joseph E., *Making Globalization Work*, W.W. Norton & Company, Inc., New York, NY, 2006.

Tapscott, Don & Anthony D. Williams, *Wikinomics*, Penguin Group, New York, USA, 2006.

The Economist, *Pocket World in Figures, 2010 Edition*, Profile Books, Ltd, London, England, 2009.

Zakaria, Fareed, *The Post-American World*, Norton paperback, New York, NY, 2009.